PILGRIM ON THE GREAT BIRD CONTINENT

PILGRIM

on the

GREAT BIRD CONTINENT

*The Importance of Everything
and Other Lessons
from Darwin's Lost Notebooks*

LYANDA LYNN HAUPT

LITTLE, BROWN AND COMPANY
New York · Boston

Little, Brown and Company

Time Warner Book Group

1271 Avenue of the Americas, New York, NY 10020

Visit our Web site at www.twbookmark.com

First Edition: March 2006

Library of Congress Cataloging-in-Publication Data

Haupt, Lyanda Lynn.

Pilgrim on the great bird continent : the importance of everything and other lessons from Darwin's lost notebooks / Lyanda Lynn Haupt. — 1st. ed.

p. cm.

Includes bibliographical referenes (p.   )

ISBN-13: 978-0-316-83664-7 (hardcover)

ISBN-10: 0-316-83664-8 (hardcover)

1. Birds — South America.   2. Natural history — South America.

3. Darwin, Charles, 1809–1882.   4. Beagle Expedition (1831–1836)   I. Title.

QL689.S58H38   2006

508′.092 — dc22   2005023262

10   9   8   7   6   5   4   3   2   1

Q-MART

Book design by Iris Weinstein

Printed in the United States of America

For Tom —

hiker
biker
juggler
recycler

— *and love of my life*

*How important it is to walk along, not in haste, but slowly,
looking at everything and calling out*

Yes! No!

— MARY OLIVER

CONTENTS

PILGRIM ON THE GREAT BIRD CONTINENT

INTRODUCTION

*C*harles Darwin's work has been plumbed for details of natural selection theory and its relevance for modern science and philosophy. His life has been told in biographies that inspire admiration for his experience, his eccentricities, his intelligence, his theories. *Pilgrim on the Great Bird Continent* approaches Darwin on a much more personal level. Here, I want to consider Darwin not just as a bearer of natural selection theory but as a human who both understood and related to the natural world in such a way that this theory became possible. For as a young man embarking on his journey aboard the *Beagle,* he possessed no such vision. He developed it through work, study, persistence, and a measure of earthen grace — a grace that even Darwin, striving to be wholly scientific, admitted.

When Darwin stepped onto the HMS *Beagle* in 1831, he was only 22 years old. He had no official position on this naval vessel in service to the Crown but was invited along as a gentleman companion to the fastidious, melancholy, equally young Captain FitzRoy, and to work privately as a naturalist.

2-page map TK

2-page map TK

1-page map TK

Darwin's father paid his expenses. The voyage, originally scheduled to span two years, dragged on for nearly five. The *Beagle*'s ostensible mission was to use new marine chronometers to accurately chart the perimeter of southern South America and to resolve discrepancies on existing charts and maps. But the emerging business of colonialism drove other dimensions of such voyages. The British would use the information obtained by this and other vessels to determine naval and commercial plans, plot the exploitation of resources such as gold, diamonds, and guano (a valuable fertilizer in nineteenth-century England), and establish Anglican missions. During the *Beagle*'s many coastal moorings, Darwin stayed ashore and took overland journeys as often as possible to escape his perpetual seasickness and to study natural history, which he pursued with increasing depth and intensity.

Darwin had been attracted to the study of natural history since childhood and had always maintained a rather expansive view of the importance of his own observations. When he was not hiding under the dining room table reading *Robinson Crusoe* for inspiration, he was traipsing the grounds of the family estate, making notes about the comings and goings of the pheasants, the state of various swallows' nests, the colors of butterflies, the shapes of clouds. All these he would solemnly report to his father and sisters in the evening (Darwin's mother died when he was eight), not bothering to differentiate between the common blackbirds he actually saw and the rare birds whose presence he invented.

Charles intended to follow in his father's footsteps by studying medicine at Edinburgh University, but he was altogether repulsed by the surgical theater. The screams he witnessed during the amputation of a child's leg without anesthesia tormented him through the whole of his life. He ran from Edinburgh with his hands over his ears and knots in his stomach. Dr. Darwin was not impressed, but he helped his son to fashion an alternative career; Charles would be a clergyman and take up his studies at Cambridge.

It was here that Darwin found naturalist mentors, and he ingratiated himself so thoroughly with the brilliant botanist John Henslow that his fellow students began to refer to him teasingly as the "man who walks with Henslow." He remained a desultory student as far as his clerical preparations were concerned, but he attended natural history lectures, helped Henslow to organize his collections, joined field excursions, and — in line with the Victorian craze for natural history curios and cabinets — zealously collected beetles.

About this time, Darwin happened upon some family financial documents that confirmed what he had previously suspected: He was rich, or at least rich enough. As a member of the landed gentry, he should hold some sort of position (and clergyman — a rather amorphous and symbolic post in Victorian England — would be perfect), but he would never have to engage in the tasteless work of earning a wage. Pleased, Darwin decided to indulge his longing for travel before settling into his comfortable life happily collecting beetles into his old age while raising dogs and a few children on the side. As he dreamed over the travel memoir of his idol Alexander von Humboldt, Darwin threw himself into planning a trip to Tenerife, dragged unwitting friends into the plan, and was trying to think up a tactful, persuasive way to ask his father to fund it. But then the offer from Captain FitzRoy came, and Darwin joined the *Beagle*.

Stepping aboard the *Beagle*, Darwin appeared to be nothing out of the ordinary. He was interested in natural history and clearly displayed some aptitude. His teachers liked and

recommended him. He was delighted to have the opportunity.
And that's about it. In his letter
of recommendation for the *Beagle*
post, Professor Henslow wrote
of Darwin that he was not a
"finished naturalist." He was a
beginner, still relying heavily on
books and teachers for informa-

VERMILLION FLYCATCHERS

tion; he was not *finished*, but for this little post he would do
nicely. Many biographers have written obliquely of Darwin's
transformation — he left London unfinished, he studied all
manner of life on earth intensely for five years, he had many
adventures and brilliant thoughts, and he came back "finished," a naturalist with his life and work well in hand. In his
introduction to a recent edition of Darwin's *Beagle* journal,
Steve Jones writes, "Darwin left England as an enthusiastic
amateur, but returned as a professional scientist."

But *how* was this transformation effected? I have read accounts hundred of pages long of Darwin's voyage and studies
on the *Beagle*, and I have searched for the locus of this change.
Surely it is not this simple, from "unfinished" to "finished" as
if the *Beagle* were some sort of Victorian school of etiquette. A
biographical recounting of Darwin's activities during this
time, however detailed and exacting, cannot account for such a
turn. Darwin set foot on the *Beagle* as a shiftless youth with a
faddish passion for collecting beetles and emerged as a philosophical naturalist who could draw scientific truths from the
simple stories spun by the creatures that crossed his path.
What happened to Darwin?

In his own published writing, Darwin unwittingly perpet-
uates the mystique of his own transformation as a naturalist.

Darwin's most popular publication from this time is the *Journal of Researches*, a book he put together in the two years after he returned to London, gathering material from his diaries and natural history notes. The whole book is written in the retrospective voice of a mature traveler, and of a naturalist who has found footing in both the biological world and the world of scientific society. The book has a fine polish — Darwin's understated version of panache. His self-deprecating sense of humor is everywhere in evidence, but the "unfinished naturalist" who boarded the *Beagle* as a wide-eyed and rather aimless young man is nowhere to be found. In excavating Darwin's transformation I want to go back — back to his diary as it was written day by day, back to the notes he traced as he watched orange birds scream from tropical trees, back to his first thoughts as he confronted unfamiliar places and creatures. In these lesser-known works of more immediacy — the diary, the correspondence, the specimen notes, the unruly scribbles in his tiny pocket notebooks — there are moments that have fallen through the cracks of Darwin's more colorful adventures as they have been recorded by both his biographers and himself.

As a student of birds myself, I have always been most interested in Darwin's avian studies, and, in addition to his diary, correspondence, and general specimen notes, it is these that I have mined most thoroughly as I've attempted to make Darwin's evolution as a naturalist intelligible. When most people think of "Darwin's birds," it is the finches of the Galápagos Islands that come most readily to mind. This makes sense — the Galápagos birds were certainly important for Darwin as he honed his theory of natural selection as the mechanism of evolutionary change. And later students of evolution, some of them quite famous in their own right, returned to the islands in the generations after Darwin, studying the finches and making them even more popular.

But the Galápagos Islands were just the culmination of Darwin's five-year voyage on the HMS *Beagle*. Before reaching the islands he spent four years as a student of South America's birdlife, amassing a fine specimen collection and making observations that would fill a thick journal. These *Ornithological Notes* are little known to all but a handful of academics. They were published in 1963, eight decades after Darwin's death, in a source that most of us would consider obscure, *The Bulletin of the British Museum,* transcribed there by Lady Nora Barlow, granddaughter of Charles Darwin and a colorful scholar in her own right.

The *Ornithological Notes* form the rough basis of what would eventually be published, with hindsight and advice from experts in London, as the bird volume of the *Zoology of the Beagle*. In the *Zoology,* as in other manuscripts that Darwin prepared for publication, we find the polished results of his ornithological study and contemplation. In the *Notes,* we find the study and contemplation themselves, and they are wonderful — quirky, zealous, irreverent, and humble. In the past, Darwin's *Ornithological Notes* have been dissected for clues regarding his development as an evolutionary thinker, and they are certainly valuable in this respect. Buried in the inky scrawls is Darwin's very first mention of his own ambiguous belief in the transmutation of species. But with this singular focus, some of the equally important aspects of his avian studies are entirely overlooked.

The *Ornithological Notes* reveal not only the seeds of Darwin's thoughts on evolution but also his deep sensitivity regarding the behavioral and ecological study of animals in their natural, wild places. Here, too, something of Darwin's own eccentric, wild mind unfolds. Though such things are rarely discussed with relation to his work, Darwin's *Notes* impart calm instruction on how to watch, how to think, how to twine beauty with science, and objectivity with empathy. How to uniquely combine poetry, humor, and wisdom in the contemplation of nonhuman beings. How to deepen an understanding of oneself in the face of a wild, exuberant, natural world. How to do all of these things with wit, with joy, and with grace.

I was thrilled to discover the *Ornithological Notes* several years ago. I carry around photocopies of the *Notes* now, reading them over and over. They are so marked and notated that I can barely make them out, and because the little comments I have penciled in their margins represent so many months of thought and study, I consider this tattered heap of photocopies to be one of my most valuable possessions, which is, I admit, pitiable. Stopping the car to run into the post office, I do not hesitate to leave my purse on the front seat, but I compulsively clutch this sad pile of Xeroxes to my breast, as if they held some sort of top state secrets.

Very recently, Dr. Frank Steinheimer at the British Natural History Museum in Tring, published a much-needed list of Darwin's bird collection from the *Beagle*, referencing each specimen's modern common and scientific names. But when I came to Darwin's avian studies, and as I was writing this book, there was no such list. For this reason, I received Dr. Steinheimer's beautiful research with some cynicism and gnashing of teeth, though I suppose the timing was all for the best.

As it was, I worked with difficulty to find my bearings in Darwin's notes, gathering a great heap of ornithological arcana in the process. Darwin's own background in ornithology was limited, and many of the reference texts in the *Beagle*'s library were already outdated. The names Darwin used in the *Notes* — nearly always Latinized scientific names — are often incorrect, or very different from the ones he and the experts settled on for the *Zoology*, and most of these names are different from the ones we use today. Sometimes Darwin did not even hazard a guess but referred to a specimen as "some kind" of shorebird, or hawk, or owl, or even — at his lazy worst — simply "bird." It took me months of work, cross-referencing the *Notes* with the finished *Zoology* and referring to modern and nineteenth-century taxonomies, art work, and experts in the field, just to figure out precisely which birds Darwin was writing about. Once I'd done my best in this regard, I settled in to learn what I could from these lovely, little-known writings. To me, it is here in these obscure *Notes,* beneath all of Darwin's better-known, more widely published work, that the story of his evolution as a naturalist is best told.

Certainly I am biased. I am a seasoned watcher of birds, my eyes readily pick them out of dusty, leafy corners. In Darwin's body of work, I am drawn to birds, as we are all drawn with wider eyes to what we love. Perhaps it is artificial to pluck Darwin's *Ornithological Notes* out of his piles of work and to focus on them as the locus of his conversion from a mere insect collector to a biological visionary. Darwin thought of himself as a geologist, never an ornithologist, and he had equally developed interests in marine invertebrates, earthworms, and orchids. But birds are a perfectly natural locus for this story.

For one thing, birds are accessible to the general reader. Nearly everyone has a basic working knowledge of birds. I can mention that Darwin was observing a kind of pheasant, a bird the size of a robin, or even a specific bird, like an Andean Condor, and be tolerably certain that my meaning will be met.

I cannot say the same for discussions of marine invertebrates, reptiles, fossils of extinct mammals, or minerals — even if I could discuss these things with passable depth.

Beyond this, though, there is a freshness in Darwin's approach to birds that speaks to my purpose. When Darwin left Cambridge to board the *Beagle*, he was full of ideas — ideas put in his head by his botany instructor John Henslow, whom he adored; by the travel writer Alexander von Humboldt, whom he strove to imitate; by the imposing geologist Adam Sedgwick, whom he admired and perhaps feared. But Darwin didn't know any ornithologists, and not much about ornithology, a branch of biological science that was still in its infancy. With a minimum of preconceived ideas, without previous experience, and without expectations and intellectual constraints from his teachers, Darwin's mind was allowed to soar unfettered. It is little surprise that observations of birds would provide much of the best evidence for his theories. Over time he grew as a watcher of birds, and he was elevated in mind, in imagination, and — this is a word rarely used in connection with Darwin, but I will argue for it strongly — in spirit. I rely heavily, though not entirely, on Darwin's avian studies as I trace his conversion to a particular way of seeing — a biological vision that is relentless, patient, and steeped in a naturalist's faith that small things matter. It is a vision that has only gained in import as ecological time has passed.

I have spent much of my own life attempting to develop as a naturalist, however imperfectly. In my mind, a naturalist is someone who comes to understand the biological life and ecological relationships of a particular place with some depth and seeks to use this understanding to forge an appropriate relationship with earthly life. This is the task that grounds *Pil-*

grim on the Great Bird Continent as I follow Darwin through his voyage on the *Beagle* and then home again, as he encounters rheas, condors, owls, and hummingbirds, corals, barnacles, fossils, and pigeons. There is no single happening, single location, or single moment that we can point to, saying, "Yes, here, Darwin is now a finished naturalist." There are, rather, a string of moments, unsung moments, silences resident between moments that unfold over time, over years. Imagination serves, on occasion, as a bridge across the spaces, allowing all of these moments to be strung together, to tell a winding story.

This book is not in any way meant to pose as a biography; it is a gleaning of those instances in Darwin's life and work that inspire a renewed vision of the relationship between the human and natural worlds, and a glimpse into the various ways these older stories might mingle with newer ones. Darwin's very personal scientific methods grew out of the observations contained in his field notes, and in their creases he foists upon us his strict but beautiful maxim. Nothing in the natural world is beneath our notice — he almost whacks us on the head with it. *Nothing.* In a modern scientific era that discards heaps of organisms as unworthy of representation in a scientific journal because they lack "statistical significance," I try to take Darwin's vision to heart.

In his wise, unusual essays, farmer, poet, and conservationist Wendell Berry reminds us, "It is not quite imaginable that people will exert themselves greatly to defend creatures and places that they have dispassionately studied. It is altogether imaginable that they will greatly exert themselves to defend creatures and places that they have involved in their lives and invested their lives in." Darwin's way of seeing offers an antidote to dispassionate study. His manner of deep watchfulness allows the ordinary ground of life to become sanctified, to be brought into *sensus plenior* — a "fuller sense" — through the offering of simple attention. Such receptivity is an answer to the unnatural fantasy of human or individual isolation in relation to the rest of the natural world, and the ill health it predicates in our bodies, our spirits, and the community of life on earth. In 1830, Darwin could not have foreseen the current unraveling of ecological balance, a wild earth that requires "defense." But it is in light of these realities that Darwin's own evolution as a naturalist holds such meaning for us today — as activists, as scientists, as birdwatchers, as homespun naturalists, as everyday humans whose lives constantly brush the perimeter of a wilder, natural world.

CHAPTER ONE

Darwin's Bump of Reverence

Himantopus, legs rose pink. This bird is very numerous, in small; & sometimes in tolerably large flocks.
On the great swampy plains and fens between the Sierra
Ventana & B. Ayres. The genus has been wrongfully accused of inelegance; the appearance of one of these birds
when walking about shallow water, which appears to be its
favourite resort, is far from awkward.

— *Ornithological Notes,* MALDONADO, *May* 1833

Darwin watched Black-winged Stilts on the wide,
swampy fens of Maldonado, near Buenas Ayres on
the east coast of South America. "Legs rose pink,"
are the first words set down in his notes on the species.
Though the bird's slender body and long neck is a vivid, con-
trasty black-and-white, it is the legs of the bird that demand a
watcher's first attentions, for they are improbably long —
nearly sixteen inches long with just a few inches of bird

BLACK-WINGED STILT

perched atop. The eyes are ringed in the same lipstick-rose pink that colors the legs and bill, giving the bird a gentle, wide-eyed look. Though Darwin was sure that he was seeing a species of the genus *Himantopus*, he was not positive about its specific name. When Darwin returned to London at the close of his journey, the ornithologist John Gould helped him to determine the name current in his time, *Himantopus nigrocollis*, water bird with a black neck, which today is called *Himantopus himantopus*. The North American variant is the Black-necked Stilt, *Himantopus mexicanus*.

The *Himantopus* stilts are loud. "Their cry," Darwin scribbled in his notes describing the collective flock, "is curiously alike to a little dog giving tongue when in full chace." In the *Zoology*, Darwin's imagery is still more vivid: "When I travelled across the plains, I was more than once startled, when lying a wake at night, at the distant sound, and thought the wild Indians were coming." After one such poor night's sleep, Darwin watched the stilts some more. He saw that they gathered most often in small flocks, but sometimes in fair numbers; that they waded in shallow waters, picking invertebrates from the muddy substrate; that their necks curved and reached — long, smooth, and delicate.

Earlier writers had been taken in by the long legs and their seeming disproportion when considered alongside the species' slender body. Stilts were considered gawky and maladroit. But this is not what Darwin saw. Directly belying much better-known ornithologists, Darwin wrote that the stilt's gait is "far from awkward," that it had been "wrongfully accused of inelegance."

When I set off down the interstate, heading from my home state of Washington toward my first year of graduate studies in Colorado, I had never seen a Black-necked Stilt, and this was something of an embarrassment. I was supposed to be an active birder, and stilts actually nest in Washington State, albeit very locally and in small numbers. In 1977 drought conditions in California and Nevada, where the birds are much more common, forced the birds northward into Oregon and Washington. Irrigation of Washington's central basin had recently created a mosaic of small ponds hospitable to the stilts during their breeding season. And though I loved shorebirds above all others and would often go far out of my way to observe them, as far as stilts were concerned, I had simply never been in the right place at the right time. So when I saw a group of seven Black-necked Stilts just off the highway in central Utah, truly thrilled and astonished, I swerved, stopped, jumped out, and pulled my spotting scope from the trunk.

The species is common in Utah, but stilts were the farthest thing from my mind on that wide, rocky, golden expanse of earth, hot in the August sun in my un–air conditioned Subaru. I was trying to enjoy the astonishing landscape but was hopelessly distracted and believed I was melting. A child of northwest forests, I was a complete heat wimp. I had rigged a beach towel to cover the sunroof, trying to get some shade, and it kept falling on my head. I was hungry, headachy, and pathetic. Then, suddenly and incomprehensibly, there were the stilts, lifting their thin red legs up and down around a tiny, mysterious puddle. I couldn't figure this puddle out. It didn't seem deep enough to survive five more minutes of the sun without evaporating into the ether. And it was surrounded by a small but depressing heap of garbage. All of it was incongruous in the midst of this rocky world that, except for the birds, was entirely the same color of gold.

This, I was sure even at the time, was a know-nothing attitude. Surely the gold surrounding me must not be simply gold but shades of red, brown, and other colors with names I do not even know. Just as natives of the far north have several names for snow, and just as I, incredulous at a friend from the East Coast who, on a walk through the Olympic rain forest of coastal Washington, said to me "All these birds are boring, and brown," raised my eyes in mock horror, asking "Boring? Brown?" Good heavens, let us forget the extravagant and obvious example of the yellow-black-and-white Townsends' Warbler and consider the beautifully subtle Winter Wren, Fox Sparrow, Swainson's Thrush, Spotted Owl, and Marbled Murrelet, all arrayed in the splendor of fawn, beige, taupe, olive, brick, tawny, ocher, and cedar, all twining seamlessly with the soft damp of their enormous surroundings.

So I saw stilts — startling white, black, and rose against bright, hot gold — and I watched them, smiling, with sweaty wetness covering my body and running down the center of my back. Though the shoulder of the highway was wide and I was far out of the way of traffic, cars and semitrucks honked noisily at me, as if they could not comprehend such stupidity. A freckled waif standing on the side of the road, watching a heap of litter with a telescope. I took a lesson from the stilts, who were entirely unfazed, stepping neatly over the garbage, lifting glittery wet feet out of their round, simple puddle. Soon I entirely forgot the cars, my nearly empty gas tank, the tickle of sweat, and even the wild gold of Utah. One of the stilts looked straight at me. "He was right," I whispered, "you *are* elegant."

These days it is common for writers to speak of the long-necked waders as possessing a certain grace. But Darwin's mentors did not see the birds this way, and for his time Darwin's outlook was fresh, and full of poetry. Once he got his bearings on the *Beagle*, moving beyond his initial tentantiveness and beginning a new life as a naturalist in his own right, Darwin was never afraid to step beyond, or even completely trample, anything that had been written before him. Fifty years later he would list this tendency as one of his most favorable personal characteristics. "I am not apt," he wrote in his slender memoir, "to follow blindly the lead of other men."

Where his teachers saw a plover perched unceremoniously upon too-long legs, why did the young Darwin see elegance? Why, I could just as easily ask, did Darwin come to see evolution by natural selection where the best scientists of his time saw a heap of mockingbird specimens, a mess of organisms covering the earth, living discrete, unconnected lives? The first question is simpler, but I am inclined to believe that the answers are alike.

Darwin first observed Black-winged Stilts in May 1833, one and a half years into his five-year voyage on the HMS *Beagle*. The bird he killed for his scientific collection at Maldonado was specimen number 1,221 in a running list that included rocks, insects, fossils, marine invertebrates, birds, fish, reptiles, and some mammals. It was the sixty-ninth bird skin Darwin had prepared. Each specimen was carefully preserved, named to the best of his ability at the time, described with as much detail as his observations would allow, and pondered in the recesses of his seasick head. And though in his continuing conversion as a naturalist, Darwin came to rely above all on observation of the living animal, he would eventually amass and describe over five thousand specimens. He was careful, attentive, rigorous, and seemed truly born to this kind of work. Still, by the time he packed away his stilt skin to be shipped to London for further study, Darwin was impressed, perhaps even stupefied, by his own industry. And so was everyone else.

When Charles presented his rather imposing father with his opportunity to join the HMS *Beagle* as a naturalist on its journey around the world, Dr. Robert Darwin was unimpressed. It merely reinforced his belief, only further validated recently by Charles's shunning of the medical career he had been groomed for, that the desultory youth was "interested in nothing but shooting and dogs." (It wasn't entirely fair, of course — young Charles was also an accomplished beetle collector!) He had essentially squandered his time in medical school at Edinburgh. He had transferred to Cambridge to prepare for the clergy — a respectable occupation for a member of the landed gentry and one that would afford Darwin plenty of time for naturalist endeavors and the comfortable raising of a family. At Cambridge Darwin pursued his natural history interests and made a mentor of the brilliant botanist John Henslow. His grades were admirable, but he remained intellectually indifferent to studies outside the natural sciences. It is little wonder that his father was concerned.

"You will be a disgrace to yourself and all your family," Robert Darwin told the son who would one day be royally entombed at Westminster Abbey. The doctor composed an imaginative list of objections, amusing in hindsight, that, for all their disapproval, betrayed a sincere affection for Charles. The voyage was a "wild scheme," a "useless undertaking," and "disreputable" to Charles's character as a clergyman. Not only that, but being on a ship was just like being in "gaol," "with the disadvantage of being drowned."

Charles Darwin was a respectful son (and besides, he needed Dr. Darwin's funding to make the trip), and shattered by his father's lack of support, he immediately composed a letter refusing the post. Darwin's uncle Josiah Wedgwood, whose famous porcelain graced the table of the queen, was much more jovial and down-to-earth than the doctor. Charles took his father's list of objections to his uncle, and they went through it together, Josiah writing his own letter refuting the list point by point. The clincher was rather philosophical. Certainly the voyage *would be* "useless as regards his profession," Uncle Jos wrote to his brother-in-law, "but looking upon him as a man of enlarged curiosity, it affords him such an opportunity of seeing men and things as happens to few." The doctor, respectful of Josiah's common sense and perhaps aware of his own tendency toward overreaction and rigidity, relented. I doubt that he stopped nursing his certainty of Darwin's ultimate failure, but at least he managed to mask his concerns with doctorly advice: "For seasickness, eat raisins."

Darwin boarded the *Beagle* at the invitation of Captain FitzRoy, who was hoping to find some "well-educated and scientific person" who might offer insight into the geological and biological dimensions of the journey. Though he was only twenty-six years old, and though he suffered severely from the manic depression that seemed to be a family trait, FitzRoy was an accomplished sailor. Darwin's rather aimless studies at Edinburgh and Cambridge contrasted sharply with FitzRoy's fastidious industry. Even so, Darwin came strongly recommended by the Cambridge network, and FitzRoy was desperate for a companion, one whom he might relate to as a gentleman rather than as a naval officer. Such a person might alleviate some of the loneliness that inhabits a naval captain's post. FitzRoy was well aware of the suicide in his family history — in the throes of a deep depression his maternal uncle had slit his own throat. Even harder to dismiss was the unshakable image of the *Beagle's* previous captain, who, unable to withstand the stress and loneliness of running a ship for years at sea, stood in the captain's quarters and shot himself in the head. I imagine FitzRoy stepping into that cabin alone after taking possession of the *Beagle*. It had been scrubbed and fitted with gleaming new mahogany. Still, he is like a young Lady Macbeth, rubbing compulsively at blood that is not there but that refuses to disappear. FitzRoy moved the captain's quarters to the other end of the ship.

In their meetings before the voyage, Darwin and FitzRoy generally thought well of each other. Darwin was uncharacteristically enthusiastic, perhaps in an effort to allay his father's fears. "Cap FitzRoy is in town and I have seen him," Darwin wrote in a letter home. "It is no use attempting to praise him as much as I feel inclined to do, for you would not believe me. — One thing I am certain of nothing could be more open and kind than he was to me." Darwin had no idea that although FitzRoy quite liked him, he almost dismissed Darwin as his choice companion because of the shape of his nose.

Phrenology, the belief in a relationship between a person's character and the morphology of the skull, was trendy in the nineteenth century, rather like astrology in the 1970s. Everybody had heard of it, most knew a little something about it, many were very knowledgeable, and a few were entirely taken in by it, planning their lives and choosing their friends according to the various bumps on their heads. Like most in his social milieu, Darwin was in the "knew a little something about it" group, but even as a young man he possessed a no-nonsense sort of intelligence that kept his bemused awareness from progressing into any deeper kind of interest or belief. Captain FitzRoy, though, was a bit of a phrenological aficionado.

Darwin's nose had always been quite bulbous. He was well aware of this, had never liked his nose, and considered it his least attractive physical feature. (In his younger days, this was probably an accurate assessment, but as he grew older and his brow bulged farther out over his eyes, the entire shape of his face was disconcerting. Eventually he grew a nice soft beard, which balanced things out.) At their first dinner, Captain FitzRoy, eager to approve of Darwin, could not help taking surreptitious note of Darwin's head shape in general, and his nose in particular. Large, soft, a bit drooped — *this was a lazy nose.* And as any phrenologist knew, the nose was an indicator of the capacity for personal industry and overall sturdiness of character. In light of the arduous journey that lay ahead, Darwin's nose boded ill.

But Mr. Darwin seemed such a good, upstanding fellow! From a decent family, and clearly a gentleman! A rare chap who could share the young captain's love of geology *and* Jane Austen! The departure date was approaching, and three other potential companions had turned the position down. FitzRoy would overlook the nose and hope for the best.

In the first weeks of the journey he surely wondered over his choice. Darwin had no sea legs whatsoever, and he required instruction on the smallest of details. He couldn't even get into his hammock without help, and once he did get in, he scarcely got out again, racked as he was by the headaches and nausea of perpetual seasickness. Darwin was too sick even to drag himself up on deck to see the first bit of land approached by the *Beagle*, the shores of Tenerife, one of the places he most longed to visit. He confessed to his father, "The misery I endured from sea-sickness is far far beyond what I ever guessed at." Poor Darwin lay in his hammock day and night, nibbling raisins.

In five years, Darwin never overcame his seasickness. But by the time he was studying stilts in Maldonado, he had learned to make the most of his time on land (only eighteen months of the five-year voyage were actually spent at sea), observing wildlife and collecting specimens for further study on board the ship and later back in London. And in spite of the limitations implied by his ponderous nose, Charles Darwin worked like a man possessed. He amassed and analyzed specimens. He produced heaps of writing — literally thousands of pages of journals, notes, and descriptions of land, weather, humans, animals, and orchids. But more than any of this, Darwin's work consisted of watching. With astonishing presumption, this young, educated Englishman with a trust fund got down on his hands and knees, dragged his rich British flannel through the mud, and watched the earth's creatures with an attention wholly new to the science of his time, an attention both subtle and brash. And it is in this light — as a rare watcher of wild things — that I like to consider another phrenological examination of Darwin. Years later, when he had become a renowned scientific thinker, a group of German phrenologists observed Darwin and proclaimed that he would have made a remarkable clergyman. "His bump of reverence," they noted solemnly, "is ten times the natural size."

This bump, sometimes called the "Veneration" by phrenologists, is located at the very crown of the skull, the spot that is soft on a baby's head. A dip at that location implies a very weak Veneration, and a protrusion indicates a strong one.

Most of our crowns are rather level, but Charles Darwin's was clearly bumped. "The mental activity manifesting through Veneration brings into being reverence for all forms of life, and the two great virtues of patience and humility," writes a contemporary phrenologist. He continues: "Veneration describes the higher aspirations of a subject, the quest beyond terrestrial existence." What an odd thing to think in relation to Darwin, a man who spent his life with his nose pressed closely against terrestrial existence.

I do not doubt the size of Darwin's bump, nor do I believe that he missed his calling. Rather, he applied his oversized bump to his naturalist's endeavors and, as a young man hardly knowing what he was about or what he might be doing, Darwin began his life's work, changing the relationship between humans and the rest of earthly creation forever. Darwin is considered to be one of the fathers of modern science, and certainly biologists today recognize their debt to him as they continue to debate the details of evolutionary theory. But the *way* that Darwin studied — his methods and the quality of his watching — bears little resemblance to modern scientific practice. By the 1920s, the naturalist tradition that Darwin helped to shape was replaced, almost completely, by the experimental method. Today, there is little room for naturalizing in biological science, except in the realm of the amateur hobbyist. Darwin's writing embraced all manner of things that are strictly disallowed in modern scientific journals — affection, mystery, anecdote, subjectivity, and a word rarely used with reference to Charles Darwin, but I do not hesitate to suggest it, reverence.

Reverence notwithstanding, Darwin was a product of his time, and the *science* of his time. It is difficult for modern sensibilities, certainly, to see in a man's participation in the pillage of South American birdlife a kind of reverence for life. But we have to understand that this difficulty stems in large part from the ability we have today to observe and study animals without pulling them out of the trees — a luxury far beyond Darwin's time. It is a harsh and occasionally repellent reality for a lover of birds that nearly every description in the *Zoology* begins "I shot my specimen at . . ." and very often proceeds onward to describe other affectionate acts: "I opened the stomachs of several, and found . . ." The same man who captures my heart by dropping all scientific hubris to speak of the Peruvian Ovenbirds as "our little friends" manages to break the spell utterly when he whispers sweet nothings about the ungainly, but innocent, Falkland Steamer Duck. "Their heads," Darwin writes, "are remarkably strong, so much so that actually I had difficulty in breaking it with my geological hammer."

I will return to this subject in some detail, but I want to say here that there were few alternatives of the sort that naturalists and scientists take for granted today — no portable photographic equipment and no bloodwork or DNA studies. There were not even decent binoculars and scopes to bring the birds in close for observation, or field guides to help identify them. It would have been not only unheard of but actually impossible for Darwin to work as a naturalist in the early nineteenth century and not rely on specimens. But it was the awakening eyes he cast upon his specimens, and — more instructively for the young Darwin — the living animal, that allowed him to transcend these expectations.

Darwin tried to collect — a euphemism for shoot and bring home — a male and female of each species he observed and also to examine the stomachs of one or two others in order to determine their diet. In this regard, he was at the conservative end of the emerging trend to collect a number of individuals in order to evaluate variation within species and to avoid being thrown off by a particular bird's potentially anomalous traits.

At first, Darwin had few qualms about collecting, but near the end of the voyage, he began to become more and more reluctant. Darwin's evolution as a naturalist taught him to value the live animal in its place over and above the bird in the specimen drawer and to lose all relish for the act of personally collecting.

Darwin's own methods bear witness to the living dimension of the science that he was creating. One day in Bahia Blanca, Darwin was seeking a little bird that he rightly believed to be a kind of wren. Eventually he would obtain specimens of the bird and observe it closely on the Falkland Islands, where these birds were quite tame. But on this day it was a new bird for him, and, like any birder, Darwin was anxious to see it, to see if he might be able to match it to a species that had already been named, and — always his hope — to learn something entirely new about it. At the time, the bird was called *Troglodytes platensis* by scientists — "wren of the plains" — and today it is commonly known as the Sedge Wren. Once Darwin obtained a specimen, he was able to identify it correctly according to the nomenclature of his time.

Before that, however, he had only heard the wren sing and caught a glimpse of its lifting brown tail. The Sedge Wren is a shy bird, and it tends to disappear utterly into the reeds. "Frequently having marked one down to within a few yards in the open plain," he complained in the *Notes*, "I could by no means obtain another glimpse of it." The bird lives "close to the ground, in the course grass which springs from the peaty soil. I do not think," Darwin mused in the *Zoology*, "I ever saw a bird which, when it chose to remain concealed, was so difficult to disturb." But he tried. He walked back and forth and back again over the place he was sure a wren had nestled, hoping to "obtain another sight of it."

He spent more than four hours one day, sometimes watching, sometimes calling, sometimes silent and still on his knees in the mud, waiting with patience and sure expectation for the appearance of the little bird. It seems he loved the bird and its precious smallness all the more for its secrecy, its sturdy hold on its own tiny bird life. Dusk and dinnertime arrived sans bird, and finally Darwin stood up, newly aware of the clumsy presence of his body, a body he had all but forgotten in his attentive watching. He scratched the fresh red insect bites on his neck and wrists. He shook his head to bring himself back to the enormous world and to the realization that would not leave him: We are unbelievably small, we are highly forgettable.

And it is this realization that separates Darwin's work from other scientists' of his time — and of ours. He was coming to understand the completely unique value of the individual while at the same time realizing that it is a mistake — a kind of hubris — to confer a more transcendent sense of import to any individual animal, whether that animal is a sea star, a wren, or a human (and his failure to set humans apart, of course, got him into deep trouble with conservative Victorians after the publication of the *Origin of Species*). The art historian Sister Wendy Beckett links this kind of understanding to the central definition of reverence. I picture her whispering lispily in some fresco-filled chamber, "Reverence is the heart's deepest form of respect, of love. It accepts that we as individuals are not essential to the universe." The individual is beautiful, worthy of awe *as an individual* — and is, nevertheless, utterly dispensable. It is a truth that defies logic, a paradox of a truth, but one that underlies and defines Darwin's understanding, one that will eventually allow him the insight required by his evolutionary theories. Good ecologists understand this today on some level; only in an unhealthy ecosystem does the individual gain a distorted significance.

In the Pacific Northwest, where I live, the last Gray Wolf died in 1919, killed for a bounty. It lived until then in the damp, green, riverine forests of the Dungeness Valley in the coastal Olympic range. The valley is drier than the famous rain forests another sixty miles north but still lush, with big red cedars, Douglas firs, and native rhododendrons with pink blossoms, rhododendrons that crawl with orange native honeysuckle. The rocks in the Dungeness River are silver-gray, like the wolves were. There used to be wolves here — families of wolves walking the forests and drinking from the rivers. I find myself too weak in spirit to spend much time imagining their wide, wild lives, their lanky legs and big tails. It is not one hundred years since they have been gone, and yet they are gone so completely.

But lately a wolf was seen in Washington State. She wore a radio collar and bore a number, so we knew exactly which wolf it was — one that had recently been reintroduced in central Idaho and had wandered. She made her way back to Idaho, and scientists tracked her the whole time, carrying her collar back across state lines. It has come to this, come to a day when we must insufferably number the wildest things. The immensity of the sadness and the sickness this truth reflects is beyond words. In a balanced world, the wolves would be present beyond number. As individuals, wolves have lost their precious lack of value. Reading Darwin's notes I see it again plainly: An understanding that twines comprehension of an individual's "forgettableness" with its unique worth is the source of a naturalist's paradoxical reverence.

Darwin knew this — or rather he learned it as he continued, over years of time and over wide spaces, to watch and study. He watched birds, paying close attention to their lives, their singular places, the color and turn and movement of their many feathers, the way they looked straight into his eyes, or did not. And he esteemed secrecy. He knew that he could watch a common wren until the sun went down, day after day, and still not know everything about it. He could lie on the ground in his almost warm enough sleeping bundle at night, making up ornithological theories until the Black-necked Stilts' rattling calls took him straight out of himself. They reminded him of what he already knew — that his theories must proceed from his watching of wild things rather than be imposed upon them. In a recent work Wendell Berry unraveled the word *theory*. "Theory," he wrote, is "at root related to the word 'theater,' it has to do with watching, with observation." Darwin was bold, very bold, in his theorizing, and had great trust in his ability to understand what could be known of the natural world. But his work does not suffer from the common narcissistic tendency to treat life — or even a knowledge of life — as though it were in his control. He was always aware of two different kinds of questions — those he did not know the answer to and those he *could not* know the answer to. These latter he considered, sensibly and to the end of his days, as beyond the scope of the scientific endeavor. Science cannot encompass everything, and in Darwin there remained room for mystery and a tolerance for uncertainty and ambiguity, even as he sought appropriate knowledge.

Darwin could see elegance in a long-legged wader because

he watched it — one single life in a particular place — with curiosity and emerging affection but without expectation. Darwin did not learn to do this at Cambridge. He learned it himself, on the rocks, in the face of wild, feathered things. He learned it on the path of his conversion from a collector who could name things to a naturalist who could see things, a conversion effected by the muse of the wild earth. No wonder his bump of reverence grew so large. When Charles returned after his five-year voyage to his father's home, "The Mount," in Shrewsbury, he walked in unannounced at breakfast. His shocked and thrilled sisters jumped up to smother him in kisses, while the reserved Dr. Darwin regarded him with quiet affection. He shook his son's hand warmly. "Your head," he said, "is quite altered."

CHAPTER TWO

Watching for Voices

From their squatting habits, they often rise unexpect-
edly close to a person. — When a pair are together,
one may be shot, without the other rising. — The whole
flock always rises together, & each bird utters a squeak like
a snipe. — From their long scapulars, when on the wing,
they fly just like snipes. — Hence all the Sportsmen of
the Beagle called them "short-billed snipes." . . . The contents of the intestines and dung is of a very bright green
colour.

— *Ornithological Notes*, ST. JAGO, 1832

When Darwin and his nascent bump boarded the
Beagle, he had very little idea what to do with himself. He was the only person on board without an
official mission. He hoped to be viewed as a competent ship's
naturalist, but he had never known a ship's naturalist. He had
never even been on a ship. And though he felt rather over-

whelmed and lost in this new situation, there is one thing he knew very well — that this was a rare opportunity. Certainly he could not have comprehended the scope of his potential impact at the time, and if he had, he would likely have fainted dead away in fright. But he knew this was a chance for him to accomplish *something*, and he didn't need a nose-gazing phrenologist to tell him he was prone to torpor. Just one week before sailing, this is what he wrote: "If I have not energy enough to make myself steadily industrious during the voyage, how great & uncommon an opportunity of improving myself shall I throw away. — May this never for one moment escape my mind." I am charmed by this simple, vulnerable recognition — that he might fail, and that if he did, it would not be a surprise.

Always susceptible to nervous symptoms, Darwin worried himself into a bout of heart palpitations, and an ugly rash broke out on the back of both hands. Convinced that his condition was surely fatal, Charles avoided doctors, fearing he would not be allowed to sail.

The young Darwin hoped against hope that he would overcome both his dissolute tendencies and his hypochondriacal heart disease in order to live up to his own fuzzy aspirations for himself. For now, they were simple. He wanted to make Professor Henslow proud of him by doing good naturalist's work, perhaps by bringing home a new specimen of bird or fossil reptile. He wanted his father and sisters to be a little impressed with him, to think of him as a *man* in general and as a *man of science* in particular. And for starters, but perhaps the most difficult challenge, he wanted to stroll around the boat without the experienced crew thinking he was a complete neophyte.

This last aspiration was far from easy. Darwin had no experience at sea whatsoever. The *Beagle* was a trim little ship, what was called a ten-gun brig, and the meticulous captain delayed the departure date in order to have it completely scrubbed and refitted with the best wood and brass before the voyage. Seventy-four people would have to pack themselves into the ninety-foot boat. Space was tight, and Darwin had done a lot of shopping: pistols, binoculars, jars and preservatives, microscope, magnifying glass, shirts, notebooks, pens, a small library. Darwin was, in fact, a gleeful shopper. "I could spend money on the very moon," he would write home from the Andes. He had always loved his own version of toys — scientific tools and gadgets. Dr. Darwin had been sweet-talked into funding a veritable chemical lab for Charles and his older brother, Erasmus, in their teen years. Now, a couple of weeks before departure, Darwin boarded the vessel with his brimming bags and gazed about his tiny shared cabin, fretting. "Went on board and returned in a panic on the old subject of want of room," he wrote neatly in his fresh new diary. "Returned to the vessel with Cap FitzRoy, who is such an effectual and goodnatured contriver that the very drawers enlarge on his appearance and all difficulties smooth away." Darwin was rather in awe of the captain.

Along with the captain and crew, Darwin spent several nights on board while the *Beagle* was still docked at Devonport. "In the morning the ship rolled a good deal, but I did not feel uncomfortable; this gives me great hopes of escaping sea sickness," Darwin wrote with sadly misguided optimism. (Just four days later, still moored, he would write, "Such a night I never passed, on every side nothing but misery.") Finally, after spending all of December anchored at Devonport, the *Beagle* set sail for the Canary and then the Cape Verde Islands in early January of 1832. When Darwin realized his seasickness was not going to abate, he finally emerged — pale, nauseated, and unsure. All the trappings were at the ready, but still Darwin wondered, just what does a ship's naturalist do?

In an archetypal human moment, queasy Darwin kept his head and managed a bit of thought. Surely a real ship's naturalist would be plucking things out of the sea and looking at them. Just ten days into the voyage, Darwin contrived a little net, four feet deep and made of bunting, to drag behind the boat, thinking it would "afford me many hours of amusement & work." Darwin's is, in fact, only the second plankton net known to have been deployed in naturalist explorations. It is likely that he had heard of the first, created by another explorer just a few years earlier, in a natural history class at Edinburgh. He was so pleased with the net that he made a small sketch of it in his diary (and like most of the drawings he would attempt in his various notebooks, this one had little to recommend it).

Other than amusing himself with his net, Darwin didn't really have much to do, so he made the most of his time. "I am quite tired having worked all day at the produce of my net." Tiring though it seems to have been for him, young Darwin was taken with the loveliness of the invertebrate animals he brought up, and he was already beginning to wonder, albeit simply, how these various beings fit into a broader natural scheme. "Many of these creatures so low in the scale of nature are most exquisite in their forms & rich colours. — It creates a feeling of wonder that so much beauty should be apparently created for such little purpose." It is amusing to find such non-Darwinian sentiments emerging from the mouth of the man himself.

Darwin reached into his store of shiny new jars and carefully preserved the several invertebrates, handwriting labels with India ink and pasting them to the glass. He opened a new ledger and next to the date January 11 penned a simple description of his first specimen, a very common hydroid known as the *Velella*, or "By-the-wind Sailor." These paper-thin disks with a purplish tint can be found by the thousands, dried up on both the Atlantic and Pacific shorelines of North America after a harsh spring wind. "A beautiful little animal," Darwin wrote next to his unhelpful drawing. That same day, the net brought forth a specimen of the more complex group *Medusae*, "a transparent membranous bag." Then, haplessly, a Portuguese Man-o-war. "Getting some of the slime on my finger from the filaments it gave considerable pain," he recorded with faux objectivity. All three were pressed enthusiastically into jars and doused in preserving spirits. Later, Darwin would pack these jars away and ship them to London, but at the time, he lined them up on his cabin shelf and looked at them for a long while, a small, Mona Lisa smile on his lips. Now he was a real ship's naturalist, and here were the specimens to prove it, all in a row.

It is a common misconception that Darwin was somehow the official ship's naturalist on the *Beagle* voyage. Certainly Darwin conferred this status upon himself in his own mind and did what he could to perpetuate the notion among his friends and family who couldn't have known better. But by rights and tradition, the naturalist position belonged to the ship's senior surgeon, Robert McCormick. He and Darwin were suspicious of each other, made catty comments behind each other's back, and were barely cordial when their paths crossed. Captain FitzRoy displayed a remarkable indulgence toward Darwin's naturalist efforts and supported him beyond reasonable expectation, while little notice at all was taken of McCormick, who was left to languish on his own. Very early in the journey, the situation came to a head for McCormick, and he huffed off the ship, returning to London and to a career that would have fallen entirely beneath our notice were it not for a Darwinian footnote. In a letter home to his sister Caroline, Charles scribbled laconically, "He is no loss." Darwin was now free to establish himself as the ship's natural history expert, unfettered.

But jars of dead things were just one part of his new naturalist duties, and Darwin knew this. A man of science must also keep a diary. Darwin began his on October 24 in Devonport, where he hoped he would be moored for just two or three weeks before the *Beagle*'s departure. The clean first page of a new diary is always daunting, even for an experienced writer. I imagine Darwin sitting with his pen poised a bit stiffly for some minutes before managing his lackluster first sentence: "Arrived here in the evening after a pleasant drive from London."

I am using the word *diary* here in a particular way. Certainly this was not the navel-gazing reservoir of daydreams, personal philosophy, spiritual reflection, and "connection with the inner child" that the self-help manuals of today encourage us to keep (though if read between the lines, Darwin's diary might be said to contain all of these things). It was a nineteenth-century gentleman's diary of a scientific journey, so some private thoughts and details would be appropriate but nothing gushy or wrenching or terribly personal. These sentiments, when they emerged at all, would be reserved for letters. (And they did emerge. "I would say poor dear Fanny till I fell to sleep," he whined to his sister Caroline upon learning that Fanny Woodhouse, to whom he considered himself all but betrothed, had married another man with alarming efficiency after his voyage began.) Although some natural history observations would be sprinkled throughout the diary, the bulk of them would be recorded in separate notebooks kept with more scientific rigor. The diary would be a repository for the journey's day-to-day happenings and travails.

Darwin himself rarely called it a diary. It was, rather, his "commonplace journal," and perhaps *journal* is a better word for it, since we so often associate the word *diary* with a kind of deep privacy that did not interest Darwin — at least not here. But because Darwin's published account of the voyage was titled the *Journal of Researches*, referring to the daily diary as a journal could invite confusion. So I am following Randal Darwin Keynes, who brilliantly edited the diary for publication and called it *Charles Darwin's Beagle Diary* for this very reason.

Darwin wrote on piles of faintly lined paper with a loosely sewn binding and red margin lines. He had the rudiments of diarizing down nicely. A mark of his earnestness is the fact that in 751 pages he never strayed from the layout he adopted on day one. The year and the geographical location appear as running headers at the top of each page. Each entry begins with the day of the month, and every fresh page begins with the name of the month as well. Each page is numbered in the top left corner. Darwin's handwriting is artless but perfectly readable in spite of his protestations. And, given the fact that the diary contains an astonishing amount of writing with a relatively clear narrative, there is very little revision. Darwin rarely crosses out a word. In the surviving manuscript are several pencil revisions, but these were apparently added when he was mining the diary for material to be published in the *Journal of Researches*, after the voyage was finished.

From the start, Darwin saw the diary as one of his most important tools. He didn't come out and say it, but he always hoped to write a travel account, like that of the German explorer Alexander von Humboldt, and the diary would be his source material. Even if that never happened, he was sure it would be of interest to him as he aged, and perhaps to his eventual children. Although Darwin never meant the diary itself to be published as written, he did assume it would be enjoyed by his immediate family and perhaps a few close friends soon after his return, so he wrote with them in mind. This imagined audience made him self-conscious — both stiff and overeager — as he sought his own writer's voice.

At the beginning anything tropical or exotic was worthy of affected notation. "The day has been very hot: & I have feasted on Tamarinds and a profusion of oranges. — for dinner I had Barrow Cooter for fish & sweet potatoes for vegetable." The meal, he pointed out in case it was lost on anyone, was "quite tropical and correct." Darwin loved to indulge the romance of tossing about naval terms as if they came naturally to him. Blithely, he wrote home about "riggings" and "maneuvers."

Eventually Darwin would settle in. Though his circumstances were fantastic, he would begin to claim a home within them, and the language of his diary would come to reflect this maturation. But for now, he was a confused and sometimes contradictory coil of homesickness, rapture, incredulity, and uncertainty. Thinking about his comfortable life at Cambridge and missing the hours of pleasant conversation over sherry with his friends and mentors, he pondered the voyage that had only just begun and wrote to Henslow, "I know not how I shall be able to endure it." Mercifully, he did not yet know that the voyage would last two years longer than he assumed. Creating himself daily in his writing and work — allowing these parallel tasks to form him as a naturalist — would become the key to his endurance.

Modern readers might find Darwin's diary, at least at first glance, to be disappointingly unreflective. But when read more closely, it reveals most of what we expect and hope from the diary form. It is, like most diaries, a rehearsal of personal ambitions, a form of self-invention, a training ground for sustained creativity, and a kind of vessel between worlds carrying his voice from a purely private place into a wider forum — his thoughts for himself made tangible in an alchemy of mind, life, pen, and paper.

The diary began to reveal to Darwin the fact that, if truth be told, his writing style was in desperate need of help. He rightly felt that he was alternately groping and boring. He began to despair. Just as he would send specimens home to various experts for examination and advice, Darwin sought expert counsel on the development of his diarist's voice. He wrote to his sister Caroline. "I have taken a fit of disgust with it & want to get it out of my sight," he told her, sending her the first five months of entries. He excused his sorrier moments, saying that he was generally very tired when writing and that he realized his writing was childish. Still, he sought her opinion, and asked with disarming simplicity: "If you speak quite sincerely, — I should be glad to have your criticism." He could not help adding, "Only recollect the above-mentioned apologies." Certainly he did not want Caroline to think that he thought the diary was good, but it is painfully clear how dearly he hoped she would approve.

Fortunately for all of us, Caroline was happy enough to offer her guidance. She hoped Charles would not think her "pert" to criticize, and it was certainly just her own judgment, but she *would* say, "I thought in the first part that you had, probably from reading so much of Humboldt, got his phraseology & occasionally made use of the kind of flowery french expressions which he uses, instead of your own simple straight forward & far more agreeable style."

Caroline's critique was astute. Darwin had been reading the travel memoir of Alexander von Humboldt almost constantly for three years morning, noon, and night in his rooms at Cambridge, after dinner at home in Shrewsbury, sick in his *Beagle* cabin, and wandering the deck before sunset. Humboldt rarely left his side, and never his mind. Humboldt's *Personal Narrative* was a rapturous, wordy tribute to his explorations of the Spanish Americas between 1799 and 1804. In the three volumes and two thousand pages, Humboldt created two nineteenth-century types — the scientific explorer and the literary travelogue. He was a Romantic, a student of Rousseau's work, and a dear friend to Goethe. He believed himself to be a pure observer, against any kind of colonialism (though any foreign exploration of the sort undertaken by Europeans during this period, including Humboldt's and Darwin's, smacked of colonialism on some level, no matter how rigorously the explorers themselves protested the notion).

Humboldt was a lofty German with a flowery French influence; Darwin was a practical, earthy Englishman, not particularly comfortable with the early Victorian flair for ornament. I would think that Darwin might have been distressed by Humboldt's overwrought use of language. But Darwin could not help himself. He was enamored of Humboldt, and his love leached into his written record.

Even now, as an adult, I catch myself falling susceptible to the style of writers I absorb in my reading hours. Many

established writers claim to stop reading when they begin a
new writing project for this very reason — for fear of being
influenced, even against their conscious intentions. I did this
far more in my young adult diaries, when I was still experi-
menting with personas as Darwin was. "He seems to be a man
of tolerable intellect," I arrogantly acknowledged of a new
philosophy professor while a stack of Jane Austen novels
graced my college night table. During the Heidegger seminar I
took during my junior year, every possible combination of
words was hyphenated, and I am sorry to report that I actually
referred, without intentional irony, to "the worldness of the
world." And what young woman would not succumb to imitating Isabelle Allende, flowering her diary with eternal sen-
tences that suggest the influence of magic and the presence of
spirits? Yes, flipping through the pages of my young diaries, I
can tell exactly what I was reading at the time it was written.

This is not necessarily a bad thing. As Saul Bellow noted,
a writer is a reader moved to emulation. Imitation is a way of
learning what suits us, and how. Unless we become trapped in
imitation, it is harmless, a means of enhancing our personal
repertoire and then moving us forward with one more angle of
vision in our bag of human tricks. One of the reasons I enjoy
finding this tendency in the young Darwin is that I recognize
it so easily, and so I smile. We might laugh over our young diaries together, he and I, and for the same reasons.

"I am at present fit only to read Humboldt," Darwin proclaimed three months into his voyage. "He like another sun il-
lumines everything I behold." And as if to prove the point, he
declared in the next paragraph, "The day has passed delight-
fully: delight is however a weak term for such transports of
pleasure . . . amongst the multitude it is hard to say what set

of objects is most striking; the general luxuriance of the vegetation bears the victory, the elegance of the grasses, the novelty of the parasitical plants, the beauty of the flowers. — the
glossy green of the foliage, all tend to this end." Darwin nearly
out-Humboldts Humboldt. "Already I can understand Hum-
boldts enthusiasm about the tropical nights, the sky is so clear
& lofty, & stars unnumerable shine so bright that like little
moons they cast their glitter on the waves."

"Like little moons they cast their glitter"? This is the sort
of writing that makes Darwin wretch when, as an older man,
he looks back on what he wrote as a youth. It is the same writing, of course, that makes this diary so sweet, so endearing. It
is the work of a young person beginning to know what he
loves but not understanding, yet, how to claim it for himself.
The writing throughout most of his first year on the *Beagle* remains self-conscious, as if Darwin knows that it is not quite
his own voice. And though he was taken a little aback by the
fact that his sisters did not droop in raptures over his words,
Caroline's incisive comments helped to effect Darwin's first,
and perhaps most necessary, shift in understanding. He really
wasn't a finished naturalist, yet. He was fumbling, and for a
time this fumbling itself would be his work, his footpath, his
life's proper tenor.

Darwin responded admirably to this newfound awareness.
He kept working, certainly, but he also kept writing. He
worked at his diary with a dedication that is almost alarming.
Anyone who has attempted to keep a diary knows that the second most difficult thing about it is beginning. The most difficult thing is continuing. Darwin confronted a pile of crisp
paper just as we all do — blank, blank, blank — strangely
capable, for an inanimate object, of staring back, and waiting.

It is as if the paper knows that it is only as valuable as you make it. To stare the pages down requires a calling up of confidence, a faith in the value of such potentially valueless work, and a slight tendency toward myopia. The young Darwin managed, entirely on his own as all diarists are, to balance all of these things.

Early on, Darwin began an interesting practice that could not have been entirely conscious. In his efforts to mature as a naturalist, he approached the diary as a naturalist would approach an organism, making the diary itself an object of study. He stared at his diary, turning it over in his mind and hands as he did the *Medusa* brought forth in his happy net. His diary is just another kind of biological specimen: Where does its truth lie? Its treasure, its center? What might he make of this squishy, overmalleable mess of his own words? In his diary, Darwin turned the microscope back on himself and watched for the emergence of his own meaning. When this meaning was not ready to reveal itself, he kept watching anyway, as if his diary itself were the elusive Sedge Wren, writing, waiting, writing some more — the weather, the scenery, the officer's idiosyncrasies, the creatures observed on a given day. Darwin liked what he wrote. But then, no, he hated it. He wrote again the next day. The fact that he did not indulge in tracing his own interior "process" in the diary itself does not make his struggle with that process any less real, or even less present on the page.

Darwin's beginning hope was that he would discover something original to add to the specimen drawers, perhaps a new fossil of some sort for the body of paleontological research. Or maybe he would write his own book about South American geological processes — and eventually he would. But it was unclear to Darwin at the start that the most original thing he would uncover was his own method, that his way of seeing and casting light on what he saw was the very thing that would allow his eventual scientific contributions. He didn't realize that he was in the thick of creating his life's most significant habit — transforming the activities of daily life into scientific insights. The parallel processes of watching his voice on paper and watching living things on earth allowed this biological vision to emerge over time. Through the interconnection of these seemingly disparate tasks, Darwin's work was formed by that wondrous twining of abandon and patience that characterizes nature's pace — a pace that suited Darwin's innate eccentricities, his bouts of wildness and his thoughtful torpor, just perfectly.

Darwin was always able to admit his shortcomings. True, he would take his self-deprecation too far now and then into the realm of actual neurotic insecurity. But for the most part he possessed a Socratic wisdom, always ready to admit what he did not know — a lesson most of us don't come to understand until we are much older, and typically only with great difficulty and in the throes of some kind of mortifying humiliation. Even writing from the Maldonado plains to Henslow, the man he hoped to impress above all others, Darwin mocked his own lack of knowledge. "There is a poor specimen of a bird," he wrote, "which to my unornithological eyes, appears to be a happy mixture of a lark pidgeon & snipe. . . . I suppose it will turn out to be some well-known bird although it has quite baffled me." It was, in fact, a very common little bird, already well known in Europe, called the Least Seed Snipe.

This was the eleventh bird Darwin recorded in his *Ornithological Notes*, referring to it as a "Guinea Fowl" and making a series of close observations, beginning the habit of watching

an animal longer than would seem to make sense to just about
anyone else. "When feeding, they walk rather slowly, with their
legs wide apart, like Quails." He continues his watching over
time: "They dust themselves in roads, or sandy places. — they
frequent particular dry spots, and day after day may be found
there." He attempts to draw near: "Upon being approached,
they lie close to the ground, & are then difficult to distinguish." He theorizes: "Some of the specimens have a black
mark like a yoke on their breast; I believe these to be the
males." (He is correct.) He considers an animal in relationship
to others: "Is the black yoke like the red Horse shoe on the
English partridge?" (It is.) He begins a lifelong habit of con-
sulting the local people with respect for their native knowledge and intelligence: "In la Plata the Spaniards call them
'Avecasina.'" And, of course, he holds the dead bird in his
hands and squints at the details: "The covering of the nostrils
is soft."

In time, he will use his own foundations to teach himself
to theorize more boldly, to watch even more closely, and —
mysteriously — to sometimes accurately identify creatures
that he has never seen or even consciously heard of before.
Other naturalists have attested to experiencing this strange
phenomenon. I have often been shocked when, seeing a bird
that I have never seen before and don't recall studying, I find its
correct name on my lips. Attention is rewarded with a stirring
of the secret knowledge possessed by our unconscious minds.

As Darwin watches for his voice, the earthbound creatures
surrounding him offer it up. His own self-expression grows
alongside his development as a naturalist, as a young person
who can understand the life of earth on his own terms. Humboldt throws him up into the stars, and he is occasionally stuck
there, quite happily. I don't want to deny him these moments

of loftiness, but I relish the instances — and they are many — when the sight of a rock, or a bird, or a new kind of fish brings him back to himself. He writes of a "gay kingfisher," or "petrels skimming the waves," almost like a normal human. He *can* write like the self he is becoming; he just has to keep practicing, writing through the imposed transcendence of Humboldt's influence to the more honest, personal transcendence located in the true, everyday things of the earth.

When we study Darwin, it is difficult not to speed through his earliest efforts toward the thing that most attracts us: his eventual theory of natural selection. Often his younger life and work are mined with just this in mind — a search for seeds of the later mature expression. But this unfinished period in Darwin's self and thinking is something to be valued and savored in itself, not dismissed, glossed over, or sifted out. I want to allow this meandering development in Darwin, just as he learned to allow it in himself, to recognize it, to let it stand on its own, and to keep from rushing it.

Darwin would eventually write a dozen books — thousands of pages — and though he could hardly be accused of skimping on words, he was never a happy writer. Writing was wrenching work for him. Still, his books are competently written, clearly argued, sometimes playful, occasionally charming, and often brilliant. "Not impressive," George Eliot famously commented upon reading the *Origin,* and while it is true that Darwin's work could not pass as literature in any artistic sense, it is undeniable that he came to write honestly, affably, and consistently. He developed a voice over time, just as he gradually developed an eye, and he claimed it. Once he did this, that voice never left him.

I dwell on Darwin's early, unformed period for two reasons: because it is unsung and because it is the foundation of his creativity. That period reveals the importance and the beauty of that which was still being formed, showing its spidery connection (someday) to the finished idea. We watch, hopefully. We keep watching. We fill our days with care, watching our words and minding our vision, and our evolution continues. We branch, we rise.

CHAPTER THREE

Pilgrim on the Great Bird Continent

I was surprised at the scarceness of birds: the extreme thickness of the vegetation seems only to suit a few tribes. — Within the Tropic the insects take a more prominent part in the animal kingdom.

— *Specimen Notes,* BAHIA, MARCH 1832

For lovers of birds, Darwin's first real avian encounter of the voyage will seem inauspicious. The *Beagle* stopped briefly at the island of St. Jago in the Cape Verde Islands, then at St. Paul's Rocks before anchoring along the Brazilian shoreline. Describing St. Paul's Rocks in a letter to his father, Darwin wrote, "It is totally barren, but is covered by hosts of birds. — they were so unused to men that we found we could kill plenty with stones & sticks." Reading about such incidents before his voyage, Darwin was certain that they were exaggerations. But anyone who has walked on a small tropical island, one that is too far from the mainland to be colonized by terrestrial mammals, knows how immovable tropical seabirds can be, sitting blithely, and with a certain stubborn commitment, upon their own tiny stretch of coralline atoll. Evolving entirely in the absence of mammalian predators, these species have no reason to fear human presence, or that of any other mammal. This fearlessness is one of the factors that made seabirds so attractive to milliners in the great feathered-hat craze of the nineteenth century. Their numbers plummeted.

The birds Darwin encountered that day were mostly gannets and noddies. Gannets are largish cream-colored birds with a strong fishing bill, black wing tips, and a soft glow of orangey feathers about the face. Noddies are terns — sleek, sinewy little birds with a long and graceful silhouette, slender bills, and neat triangular feet. Terns can be variously colored, but the ones Darwin walked among would have been charcoal, with blue-gray feathers on their crowns. Gannets and noddies are colony nesters, spreading themselves tightly across their small oceanic islets, with just scrapes, or shallow depressions, for nests. Most of them were setting eggs the day that Darwin, Captain FitzRoy, and an assortment of shipmen took a small exploring boat ashore. They filled its hulls with the bodies of birds and eggs, all of them taken in the Neanderthal-in-officer's-clothing fashion that Darwin described. They didn't kill the birds just for the fun of it, to be sure. Fresh meat was sought along the way, counted on as essential in a voyage of this sort. But in the records of FitzRoy and Darwin, it is clear that walking up to a bird and thwacking it on the head with a hammer served as a form of amusement for the men, one that would be considered inhumane today.

Darwin soon tired of hammering birds and left the filling of the boat to less refined members of the crew while he went

about feverishly naturalizing, noting the geological formations and collecting marine invertebrates and fishes. He was
reveling in his first few days ashore in the tropics, his first opportunity to step off the boat and into the voyage he had
imagined, and was rather agape at the newness of his surroundings. "It is utterly useless to say anything about the
Scenery," he wrote to Dr. Darwin. "It would be as profitable
to explain to a blind man colours, as to a person, who has not
been out of Europe, the total dissimilarity of a Tropical view."
For a young Englishman writing home to his doctor father,
Darwin was quite unleashed: "So you must excuse raptures &
those raptures badly expressed." He was delighted to leave his
plankton net behind on the ship and lift creatures from the
edges of the sea with his own hands.

Clambering about these tiny islets, Darwin thoroughly
enjoyed himself, a "child in a toy store," FitzRoy called him.
But even such tropical raptures could not have prepared Darwin for his own response to the forests of Brazil. Up to now
he had been excitable, expectant, almost dizzy with anticipation. But in Brazil he became utterly dumfounded in ways that
he expected, surely (being well prepared by Humboldt), but
also in ways that he could never have foreseen.

The *Beagle* anchored at Bahia, a northeastern coastal town
now called Salvador, on the twenty-eighth of February, 1832.
Even from the boat, Darwin was overcome by the greenness of
the land, what he called "verdure" in his diary. "It would be
difficult to imagine, before seeing the view, anything so mag-
nificent. — It requires, however, the reality of nature to make
it so." This sounds a bit overwrought, but I believe his reaction
here to be perfectly honest. One cannot imagine it — he
wants to be sure we know that — one must see it, such glowing green, such a place.

Magnificence, verdure, *and he hasn't even seen the birds yet.* I find myself wildly hopeful as the *Beagle* approaches Brazil. This is a moist coastal tropical forest, dense with broad-leaved evergreen trees, epiphytic plants, and vines hanging everywhere like hair. It is thicker, greener, moister than any place on earth, and it is filled with birds.

Ornithologists refer to South America as the "Great Bird Continent," holding as it does the highest avian diversity on the planet. According to the most recent count, there are 3,751 species of birds in South America (compared with about 1,100 in North America). They represent ninety families, and of these, twenty-eight are endemic to the Neotropics. This is the only region on earth where previously unknown avian species are still discovered with quiet regularity.

The lowland forests of coastal Brazil are particularly tangled with species unique to the tropics — manikins, cotingas, antbirds, woodcreepers, and toucanets — so I fully expected Darwin to be stunned by the birds when he set foot in Bahia. He was an Old World naturalist stepping into a New World land; the birds would look different and behave strangely when compared with the birds of Europe. They are as a whole more colorful, more variously employed, and often noisier. Darwin had seen study skins of many tropical species and should have been ready for the spectacle. Reading the diary for the first time, and reading it with my biased focus on things ornithological, I was eager for the wild rumpus to begin, for Darwin to be converted immediately to a life of avian study, to drool, exclaim, and perhaps fall on his knees. He wouldn't be the first bird-watcher to carry on in this way, not by any stretch. But what did Darwin do?

My own margin note penciled next to Darwin's April 16 diary entry from Bahia reads, "Still rambling on about all the stupid beetles." I apologize to any entomologists, but, really, it was "*coleoptera* this," and "*coleoptera* that," when strange-bodied neotropical birds he couldn't have made up in his wild imagination were watching him from the leafy branches. It all becomes quite tiresome for the ornithologically minded. When the gannet-hammering incident from St. Paul's Rocks is thrown in, the situation from a bird-watcher's perspective begins to look bleak indeed. Just how could Darwin possibly be so dreadfully slow to catch on about birds?

Not that he doesn't mention them at all. On an expedition to the Rio interior, he was charmed and amused by the hummingbirds he passed with his companions. "I counted four species," he recorded. "The wings moved so rapidly that they were scarcely visible, & so remaining stationary the little bird darted its beak into the wild flowers. — making an extraordinary buzzing noise at the same time, with its wings." (Even at this early point in Darwin's development as a naturalist, these seem to be curiously simple observations — the sort of comments a young child might make after a few minutes' watching. But recall that Darwin was fresh off the boat from England. Hummingbirds evolved in the New World tropics, and those that we see in North America originated there. English hummingbirds are linked to the Africa-Eurasian migration routes, where there are no hummingbirds. These would have been Darwin's very first, so the rapidity of their wing movement and their buzzing sound would have been, to him, entirely worthy of his naturalist's notebook.)

In coastal Rio he referred in his diary to the "number of beautiful fishing birds," and the "beautiful green parrots." In his pocket notebook he wrote that he "saw some beautiful birds: Toucans and bee eaters." *Beautiful*, in fact, was his favored and very nearly his only avian descriptor at the start, and I think it says something about his beginning attitude toward birds in general. Calling a tropical bird beautiful is rather like describing a person as nice — nondescript and subtly dismissive. Beyond their utility as food and sport, birds seem to be regarded by Darwin as an integral, if somewhat ornamental, element of his romantic tropical aesthetic. Certainly he was happy to be in the company of such lovely creatures, but they were clearly not a particular object of his naturalist pursuits. He would observe and collect them as they crossed his path, as he would all aspects of tropical nature. He would make a contribution to the tropical collections already under way in London, and he would do it with something of an understated flourish because that was his way. It was also, as he saw it, part of his duty. But he was clear that he intended to make his mark as a naturalist in the field of geology. Besides geology, he would feed his abiding love of entomology, so geological formations and insects claimed the bulk of his attention. After walking with one of the ship's officers in Rio one day, Darwin wrote, "He shot some birds & as is generally the case, I have found many interesting animals of the lower classes." Even with a bird-minded companion, Darwin's own nose was buried in the crumbling tropical soil and the brilliant array of insects it brought forth.

His specimen notes on the tropical insects were many and detailed. He dissected and collected and pondered, all with a happy furrowing of his brow. He caught a mantis and thought he had killed it by "holding for several minutes under water that was boiling, the head & thorax (to the insertion of the wings) & anterior legs." Surely that would have done it, he assumed, and in fact added, "These parts shortly were completely dead & became dry and brittle." So Darwin was a bit unsettled to find, eight days later, that the mantis was still wiggling, that "the abdomen & hinder legs continued to possess a slight degree of irritability." Belying his alarm, Darwin penned a cheerful conclusion: "This appears a well marked instance of the tenacity of life amongst insects."

On and on the entomoligcal details unfolded in his notes, while birds were still simply "beautiful." Besides honest personal interest, there was a rational choice at work here. Darwin wanted to distinguish himself scientifically, and he felt that birds had been "done." Tropical botany and ornithology were "too well known," he believed, for a young naturalist to establish himself in these fields. When we look back from the vantage point of the early twenty-first century, his decision seems a drastic miscalculation. Our knowledge of any aspect of tropical ecology was, and in most respects still is, woefully inadequate. But it is not difficult to understand Darwin's perspective. The showy collections of tropical bird skins toted home by earlier explorers were, because of their fabulous beauty and the popularity of the Victorian natural history curio cabinet, already quite commonplace in scientific circles. In spite of the fact that there were only one or two practicing ornithologists in all of England, Darwin could conceivably feel that the field was overcrowded and that he would have to uncover something shockingly original in order to leave any sort of mark in that field.

Beyond this reasoned decision to relegate ornithology to the third tier of his attention, though, Darwin made some very curious comments. He actually seems to have believed that there weren't many birds to see in Brazil, certainly not as many as he expected. "I was surprised at the scarceness of birds," he wrote after a couple of weeks in Bahia. "The extreme thickness of the vegetation seems only to suit a few tribes." And then, in Rio, in some general observations in his specimen notebook, he suggested, "After seeing a collection of Brazilian birds in a Museum; it would not easily be believed what little show they make in their native country. — Concealed in the universal mass of vegetation, the attention is not drawn to them by their notes." Brazil was a giant leafy tangle, he surmised, obscuring what few birds there actually were. He had viewed the local museum specimens, wondering, where were all these supposed birds, and why did they not show themselves?

Regarding the numbers and "tribes" of forest birds, Darwin was simply wrong. The lowland forests of the neotropics support more species of birds than any ecosystem on the entire earth. But one of the secret talents of thick, moist forests — the rain forests of the Pacific Northwest do this, too — is their ability to swallow up a shocking number of birds in pockets of shade, leaves, and stillness. It is a strange experience to step on the mossy forest floor of a temperate rain forest and only be able to guess at how many unseen brown birds are watching. Darwin expected toucans, mot-mots, manikins — glowing, colorful birds. Who would imagine that such birds could be so well hidden anywhere, even sitting motionless in the leafy shade, moving nothing but their eyelids? But they are.

Bird-watching in tropical forests requires a unique, practiced patience, one that Darwin would develop in good time but at this point still failed to possess. He didn't have "bird eyes" yet; he wasn't attuned to the subtle movement that would allow him to distinguish a bird from a quivering leaf amid the flowering parasites that draped the trees. Unless the voice was very distinctive, he couldn't pull bird songs apart from the song of the forest as a whole. He was an inexperienced bird-watcher without particular promise or motivation, preferring to blame his lack of success on a dearth of birds.

In all of Darwin's time in Brazil, he collected only one bird, which he labeled "Krotophagus." Back in England, John Gould would correct the spelling and finish the identification, which was the same then as it is today: *Crotophaga ani*, the Smooth-billed Ani, a large-billed, long-tailed, all-black bird of considerable intelligence. It is floppy, forward, not at all shy, and also quite loud. Darwin mistakenly conjectured that it might be allied to the parrot family (it is actually related to the cuckoos and roadrunners) and kindly called its prattling voice "harmonious." He seemed most interested in the bird for the contents of its innards. "In the stomach were numerous remains of various Orthopterous and some Coleopterous insects," was his only comment in the *Ornithological Notes*, managing to make his one Brazilian bird into yet another beetle study.

Within a year, Darwin's ornithological notebook would be filled with detail, morphological and behavioral secrets grasped through persistent, patient observation. The two species of large, flightless South American rheas would become one of his deepest obsessions during the whole of the *Beagle* journey. Later, the Andean Condors would take hold of his imagination, and he would risk his life to watch them fly. The famous mockingbirds and finches of the Galápagos Islands would help to unleash the mind of Darwin upon an unsuspecting world. How did Darwin get to all of this from these few, dull, unseeing observations in Brazil? I expected him to become a bird-watcher in Brazil, but by reining in my impatience and following along, I have been able to see how Darwin became something much better, something that would ground his future observations of birds and of everything else.

*　*　*

*I*f I had my druthers, I would have more friends who off-handedly use words such as *delphinium, persimmon, ephemeral,* and *verdant,* just for the words themselves, just because they are available, ready to make a sentence startling, ripe, full, and moist. In spite of the dictionary's calling them similes, *green* and *verdant* are two very different words. If the color of vegetation were capable of being represented in language in the same way that an onomatopoeia suggests the sound of a word, then certainly *verdant* would be the word. Where *green* simply invites us to think of green things — a bowl of peas, maybe — *verdant* is a foresty word wrapped in the color green, seeming to embody the earth's greenness itself.

As a poet, artist, healer, theologian, musician, and naturalist, the medieval mystic Hildegard of Bingen had her own strong feelings about this word. Eight hundred and fifty years ago, she absorbed its root in the creation of a new word — viriditas. Hildegard's *viriditas* is typically translated as "greenness" or "greening." But most scholars find something more expansive in the word, and it is clear that Hildegard was thinking of more than moss and leaves. Her lovely, revolutionary song, *O viriditas digiti Dei,* the "greenness of the finger of God," speaks to the greening quality of the divine. It embodies her sense of the spiritual condition as a state that is naturally and beautifully receptive — "moist," she says, and "warm." Like the interdependence of seed and soil, *viriditas* is an inspired co-creation among the divine, the human, and the earthly, a cultivation of earthen grace. I think again of Darwin's exclamation upon drawing near to the Brazilian coast for the first time. Ah, the "profusion of wood" whispered this young man who had not known there was so much green on the entire earth. The "verdure," he said.

On the twenty-ninth of February, for the first time in his

life, he took a solitary walk in a tropical forest, and his delight was inexpressible. "I have been wandering by myself in a Brazilian forest," he wrote, just because he could, just because he loved to see himself say it, to read it there on the page: *I have been wandering by myself in a Brazilian forest.* Wandering, myself, forest. This is perhaps the loveliest sentence in his diary to date. *Delight*, Darwin made known, was too weak a word for his "transports of pleasure." He had been reading Humboldt, yes, but he forgot it here, forgot the strain of imitation, and found a moment that was honest, simple, true, truly happy, and deeply himself. He began to calm himself, to begin seeing into things on his own. The elegance of grasses, he noted quietly, the subtle gloss of foliage, the shocking stature of the palms, and that strange forest paradox — that sound and silence can, in fact in wilderness *must,* wend about each other. A rippling stream, the woodpecker's tap, leaves brushing one another in the breeze, a distant bird's scream, even the loud ringing of insects that carries hundreds of yards, these are voices that forest silence can somehow contain. Charles stood in this forest, knowing something he had never known, had never truly imagined to exist. "Within the recesses of the forest when in the midst of it a universal stillness appears to reign."

The *Beagle* was in Bahia ten days, and Darwin spent most of that time ashore, adding "raptures to former raptures." He bundled these raptures someplace next to the new solitary stillness within him and boarded the *Beagle* for the two-week sail to Rio de Janeiro, where FitzRoy would spend nearly a month surveying the coastline. Darwin was happy for the time to stay ashore, taking a small room, collecting, and joining various odd companions on horseback journeys into the interior.

Like Bahia, Rio was characterized by rich, humid lowland forests, and into these forests Darwin again wandered alone.

His fellow travelers became used to his habit, already considered eccentric, of retreating into the wilds and rejoining the group only when he was filled.

He was awed as usual on one such walk, and his eyes were unable to settle, shifting quickly from the tangled complexity of the forest itself to its various components — startled thrush, lazy-winged butterflies in shades of blue, parasitic vines draping every branch. He touched, lightly, everything that didn't look like it could bite, he smiled just a little, and finally he reached for his pocket notebook, wanting to capture something of this strange, soft elation. "Twiners entwining twiners — tresses like hair —," he wrote quickly, as if taking dictation "— beautiful Lepidoptera —," until he was barely within the realm of words, "— Silence —," and finally exclaimed, quietly to himself and to the movement of this twining earth, "— Hosannah —."

Hosannah, he wrote mysteriously, almost without knowing it. *Hosannah* — such a strange word, not even a word in the ordinary sense but a recognition, an acknowledgment, a meeting of inner experience and outer substance. I see Darwin at this moment beginning his voyage again, not as a tourist this time but as a kind of pilgrim, living an unfiltered, firsthand experience, becoming his own witness to creation's story, putting himself, as is the pilgrim's task, within that story.

Here I almost let that most unfortunate usurpation of the word *creation* by fundamental Creationists prevent me from using the word myself. In spite of his own worries and confusion over its implications, in his books, including the *Origin of Species*, Darwin often himself used the term "creation" to encompass all of earthly life. All of the other words that we employ, hoping to avoid the mire of creationism — *earth, world, natural world, biotic life*, and even just *"life"* — are static next to the word *creation,* implying as it does an unending, creative ground of change, a matrix of constant renewal. More than this, *creation* better embodies the "nonliving" substrate of biotic life — things watery, mineral, and atmospheric. Degraded by the dull unreality of fundamental Creationism, the word is fully worth reclaiming.

Darwin referred to his notes when he wrote in his diary that evening after spending the day beneath his twining twiners. He did not repeat the mysterious *Hosannah,* which was dependent entirely on the lived moment, but he was no less overcome. He could not get over the tangle of vines, "the woody creepers, themselves covered by creepers." He spoke to that lovely sense of disorientation the forest can evoke in us, standing as we do between the extremes of treetops and soil. "If the eye is turned from the world of foliage above, to the ground, it is attracted by the extreme elegance of the leaves of numberless species of Ferns & Mimosas." This is lovely reporting in itself, but Darwin was compelled in the diary to record something of the spiritual response that had taken hold of him in the morning wilderness. Surely he could have simply pointed to the beauty of this or that living thing, but he insisted on something further as he wrote, "It is nearly impossible to give an adequate idea of the higher feelings which are excited; wonder, astonishment & sublime devotion fill & elevate the mind."

Darwin first put the words *sublime devotion* in his pocket notebook that day. Then he copied them into his diary, and, finally, he kept them in the *Journal of Researches* when it was being prepared for publication and public consumption. *Sublime* gives a sense of something rising up from below (*sub*) the *limen,* a threshold of some sort, and through this rising, this oblique passage, becoming elevated in the mind. *Devotion* suggests a fervent vow (*de* + *votus*) freely given and consecrating its object. Darwin often rewrote passages for publication, editing out, in particular, the moments when he seemed to come across as an overexcited youth. But he kept these words, *sublime devotion*, throughout his writings, committing to them strongly.

In the transition from student to pilgrim, Darwin began to overlay his original interest and enthusiasm with this new sense of felt devotion, broadening both his perception of and response to his earthly object. Years later, when Darwin's son Francis was a grown man, he would remember his father's attitude toward the simplest of organisms. "I used to like to hear him admire the beauty of a flower; it was a kind of gratitude to the flower itself and a personal love for its delicate form and colour. I seem to remember him gently touching a flower he delighted in; it was the same simple admiration that a child might feel." Reading Darwin's early comments about the Brazilian forest, I feel that I am a witness to some of those rare moments that allow humans to become the kind of beings who might bear such a response, whose scientific interest in a flower can be wide enough to include gratitude.

During the rest of his time in Rio, Darwin experienced a knee swollen beyond recognition, which he claims to have cured with heavy doses of port wine and cinnamon; he expressed revulsion over plantation slavery and the haughtiness of the Brazilian upper class in general; he "disagreeably frittered away" an entire day shopping in town; and he befriended a six-year-old girl named Theresa, applauding as she twirled and danced. There were still raptures, to be sure, and there was also further cultivation of that sense of devotion that began weeks earlier. It began to mature into an even quieter delight, a deeper sense of silence, of stillness, of repose in nature.

Again, the quick notes tell it best: "Silence well exemplified; — rippling of a brook. Lofty trees . . . the pleasure of eating my lunch on one of the rotten trees — so gloomy that only shean of light enters the profound. Tops of the trees enlumined" — *enlumined* being, I think, a perfectly useful made-up word. Again, there was the rich twining of the pure, touchable material of the earth — *brook, rotten trees, lunch, light* — with a sense of something that could not be simply summed up in the material nature of the experience — *silence, gloomy, profound, enlumined.* In his diary this same day, Darwin less colorfully described the trees as "brightly illuminated," went more deeply into his feeling "a peculiar chilling dampness," and expanded his response. "One feels an inexpressible delight," he wrote.

It is not difficult to think of this time as a kind of conversion period for Darwin — an intense turning that is the source of a new perspective. William James defined conversion as the "radical rearrangement of psychic energy around some new center of interest." This rearrangement is what allows spiritual experience — an act or feeling or encounter during which humans, in solitude, "stand in relation to whatever they consider the divine." James speaks expansively, and his sense of the spiritual matches Darwin's experience quite precisely. Darwin's own encounter with silence somehow recapitulates and consecrates his naturalist's work. Darwin could have shrugged this off, buttoned his collar, and returned to a more purely scientific residence in the natural world. Hildegard speaks to such closing off in the face of *viriditas* — a drying up that is the opposite of her moistness — a turning away from that part of ourselves that can apprehend the divine in the spreading green earth. This voluntary closing down, Hildegard taught, is what makes human lives shrivel.

But Darwin, at least at this point, did not shrivel. For Darwin's conversion was not based in some dewy-eyed, simplified "isn't nature pretty" sort of feast — the easy, pleasurable dimension that sets the parameters of the common human experience. No, Darwin's insight was whole and fierce. As much as he speaks of beauty, he speaks of decay, of rot, of damp, cold chill, of tropical parasites that cling to anything that might stand still for the shortest moment. He apprehends fully the death that grounds the forward movement of tropical life, and this informs rather than detracts from his vision. Darwin, true pilgrim that he is, puts himself in the story, stuffing food in his own mouth, tending the sustenance of his small life while perched on a log that decays beneath him, watching as his beloved dung beetles run toward his very own heap before his trousers are even buttoned.

I see Darwin on his rotting log in Brazil, and I wonder, does he sense what I imagine him to sense? Clearly he had no proper language for it. I imagine him recognizing a kind of goodness, a goodness that bears a resemblance not to the Victorian moral or aesthetic structure but rather to an earthen goodness, an expanse upon which the breadth of wildness is draped — or, no, an expanse that is this breadth itself.

I have culled here just a few of the pivotal moments, from Darwin's early journal, and I am aware that the language I have used to describe his experience is more spiritual than we usually see regarding the somber "father of evolution." My interpretation departs from the typical biographical approach, which is inclined to couch Darwin's experience entirely within the intellectual understanding of the relationship between nature and religion that was emerging in Darwin's academic milieu.

Darwin's own family was liberally Christian. In a strongly Anglican culture, Darwin's mother, Susanna Wedgwood, was, like her father, Josiah, a Unitarian, embracing what was at the time a highly intellectual response to the rituals and symbolism of the Church of England, thought by the Unitarians to be getting out of hand. Susanna Wedgwood Darwin bundled her six children off to Unitarian services every week, but when she died, their father, a liberal but indifferent Anglican, did not make any particular effort regarding his children's religious upbringing, at least not until he decided Charles ought to be a parson. But Dr. Darwin did consider his family to be Anglican, with nominal ties to the local parish, as was fitting for a man of his social status. Entering Cambridge at age eighteen, Charles Darwin was immersed in a culture of academic Anglicanism.

While at Cambridge, and even after he earned his degree, Darwin read with great native interest the works of the philosopher and theologian William Paley, particularly his 1802 opus *Natural Theology; or, Evidences of the Existence and Attributes of the Deity, Collected from the Appearances of Nature*. Here, Paley introduced the enduring "watchmaker" version of the argument for God from design, in which a wanderer finds a watch on the ground and rightly assumes that it has a maker. Modern students of philosophy are still introduced to the argument. For Paley, the design of nature was a revelation of God's own character. Given the variety of earthly life, the complex placement of organisms, and the evidence that certain animals once graced the planet and were now entirely absent, Paley eloquently argued for "special creation," an extremely popular view at the time, which suggested that God created and placed all beings on earth with foresight and perfection. Some animal and plant groups were allowed to die out and were replaced with others in an updated landscape by a very busy Creator. In spite of the popularity of French naturalist Jean-Baptiste

Lamarck's theory of evolution by adaptation and the fact that Darwin's own grandfather Erasmus had written a book about evolution (closely echoing Lamarck), young Darwin agreed with Paley that there was little evidence for this sort of development in nature, and he considered himself a de facto creationist.

Paley could construct a pretty argument, and his style — heaping terrific quantities of evidence on his theological generalizations in, basically, one long argument — would be a great influence on Darwin throughout his life. His *Origin of Species* would be written with much the same strategy. Certainly there was a prudish element to Paley's work, upholding as it did an established and constrained British morality. All things in nature, according to Paley, tended toward the good of man, particularly if man was civilized and properly dressed. Still, there was something poetic and lovely in Paley's line of reasoning, which sanctified the naturalist's endeavor by locating God in the details of earthly life. Darwin surely responded to this view in *Natural Theology.* (And I am certain that in Brazil he would have been pleased to find his own tastes corresponding quite exactly with the Creator's, who seemed to show, as biologist J. B. S. Haldane famously observed, "an inordinate fondness for beetles.")

Darwin's experience in the Brazilian wilds is often represented as a kind of religio-intellectual one, with Darwin all buttoned up in the rainforest, pondering the intricate holiness of earthly existence with deep attentiveness but at an intellectual remove. His "Hosannah" is strained, not quite ringing from the heart, his "sublime devotion" is an unwitting imitation of what he has studied.

But I am unwilling to limit Darwin's experience in this way. I myself have stood alone beneath the rain forest canopy, and I know what can happen there. I want to listen to Darwin's own hurried words, words that surprised even him, let them stand alone, and watch Darwin as he sits beneath a rain forest tree, allowing its strange insects, its leaves, its moistness, to fall right onto his hair. Why are we so squeamish? Why so prudish? Why deny Darwin the depth of his response to the astonishing breadth of the wild green earth?

Writing in 1959, historian Gertrude Himmelfarb speculated that other biographers had been "carried away by the zeal of hindsight," hurrying Darwin toward the strict materialism that many scientists like to see in his evolutionary insight. Yet it could be argued that his scientific insight might never have come to him had it not been grounded in these deeply felt experiences, had he never made the shift from traveler to pilgrim, had his sturdy intellectual sensibility not been moistened by the forest's secret teachings.

Certainly Darwin did not walk about in some sort of continuous spiritual fog. Like any experience involving expanded perception, Darwin's heightened sense of stillness, repose, and the sublime could not be sustained over time. Rather, it had to be attenuated somehow, slipped back into the slenderness of ordinary daily life. Darwin spent the remainder of his time in Brazil as a student of small things, things found at stream sides, behind leaves, and beneath stones. He was enlivened, it seems, by a new respect for that which does not show itself on the surface. He became particularly, and very patiently, interested in spiders.

Back on board the *Beagle*, about to depart, Darwin felt ready for what was next, and he was pleased, for a time, to return to the "intricacy" of his corner on the ship. He was becoming happy and comfortable in his new life, starting to settle into a naturalist's routines, observing, collecting, preserving, recording. It was from this place, from firsthand experience and the simple busy-ness of the daily round, that birds would begin to entice Darwin. Without ornithological mentors, without particular hope of ornithological fame, he would be drawn into the lives of birds. And he would come to them with a new openness, seemingly called by the birds themselves — the moist, green birds.

CHAPTER FOUR

Small Birds, Large Thoughts

A mongst the smaller birds, my collection is very perfect.

— *Ornithological Notes*, MALDONADO, 1833

I n vain may a person intently watch the thicket, whence, every now & then the noise proceeds; in vain may he try, by beating the bushes, to see its author; at other times by standing still, especially within the forest, the bird will hop close by.

— *Ornithological Notes*, MALDONADO, 1833

T here was a particular little bird that Darwin glimpsed just a handful of times. He saw it in the reedy coastal marshes at Bahia. He discovered a dusty study skin of the bird in a museum in Chile. Finally he obtained his own

specimen in Maldonado, in southern Uruguay. He didn't
write much about it in his notes, but the entry is one of my fa-
vorites, and the bird is one that clearly touched Darwin on a
different level than most. It was a year and a half into his journey. This is what he wrote:

> Parus (?), Exquisitely beautiful — very rare, frequents
> reeds near lake — soles of feet fine orange.

My first penciled note next to this entry in Darwin's *Ornithological Notes* says, "What is it?" What is this *exquisite* bird, a
word Darwin so rarely employs, with its surprising feet? As
usual, Darwin was trying to link the tiny bird to a European
genus with which he was familiar. The genus *Parus* encompasses a group of birds in the family Paridae, the tits. In
North America we are most familiar with the very cute chick-
adee and Tufted Titmouse, both in the genus *Parus*. In Europe
the genus is even larger, with several more colorful birds in
shades of black, powdery blue, white, and yellow and often
with bridled faces and more complex plumage patterns. So I
had this to go on: Darwin thought his orange-soled bird
might be a sort of Parid, and, though there are none in South
America, the bird must had resembled one of the European
varieties Darwin carried around in his brain.

John Gould's more official identification in the *Zoology* offered clues, though nothing definitive. Following the British
naturalist William Swainson, he identified the bird as *Cyanotis
omnicolor*, a name that did not survive to modern times. But at
least Gould's designation told me that the bird was multicol-
ored enough to warrant such a description in its scientific
name. One very nice thing that Gould did in the *Zoology* was to
list all of the scientific names that each bird had been given by

other ornithologists, and in this case the list helped me determine the modern name of the orange-soled bird. One French naturalist named it *Sylvia rubigastra*, and another French naturalist called it *Tachuris omnicolor*. There is, today, a *Tachuris rubigastra*, a small flycatcher, that fits Darwin's description and has a range consistent with Darwin's field observations of the bird. Its common name is the Many-colored Rush Tyrant.

Although I have watched birds in the neotropics, I have not yet ventured as far as southern South America, where Darwin traveled and where the Many-colored Rush Tyrant lives. But in the fine paintings and exegesis by Robert S. Ridgely and Guy Tudor in my cherished two volumes of *The Birds of South America*, I can glimpse its beauty and habits. It is a very small bird, barely able to contain its colors, though it wears them blithely. Locally it is called *Siete Colores*, though actually it has eight, not seven, colors (or perhaps even nine, depending on the slant of the light), with a vivid yellow breast, olive green back, orange undertail, white-and-black-striped wings, violet blue cheeks, and a crown striped with orange, black, and more yellow. There are two black half-moons crossing the breast, not quite meeting in the middle. The bill is black, and slender for catching small insects. The legs and tops of the feet are dark gray-black. And yet, in spite of published descriptions to the contrary, the bird does not, to me, appear gaudy. The colors are all arranged so neatly, so efficiently. How to fit so many colors on such a tiny bird without making a mess? Yet here they are, a study in avian economy, and the effect is one of utter simplicity and order. Those feet do seem a sudden extravagance, though. I can't find the soles of the feet in any book, but this strange fact that they are "fine orange" has captured my imagination, as it did Darwin's. Why would they be orange?

Darwin called the bird "rare," but we use that word more stringently these days, and the rush tyrant is now thought to be "locally common," meaning that if you are in just the right spot, you can expect to see it. It is very busy and conspicuous in its reedy haunts, sallying forth from its perch on upright water plants to catch insects aerially, then landing back on its stem, swinging up and down, both behaviors that Darwin was able to observe closely.

All of this coalesced for Darwin in a special bird, a bird that was "exquisitely beautiful." This use of the word is different in tone, somehow, from the *beautiful* that he used to toss off for any good-looking bird that crossed his path. The note on the Many-colored Rush Tyrant appears in the middle of several pages of fine, but more prosaic, natural history — what birds are eating and doing, and what their songs sound like. In the midst of these other notes this one comes across as singularly quiet and true, almost tinkling. Here is a beautiful creature to give us pause and rest.

Darwin's response to the Many-colored Rush-Tyrant is representative of a more expansive response to birds in general. In Maldonado his ornithology absolutely blossomed. The *Beagle* docked there at the very end of April 1833, nearly a year after his encounters with the Brazilian forests. In the intervening months, he had explored great swathes of eastern South America — Monte Video, Bahia Blanca, the Falkland Islands, and all the way down to Tierra del Fuego. In all of these places, and throughout the first eighteen months of his journey, Darwin listed only forty-eight specimens of birds in his notes. In his two months at Maldonado, he described eighty specimens and closely observed many more in the field. Before Maldonado, Darwin occasionally engaged in some fine bird-watching. His waiting for the elusive Sedge Wren occurred during his first year, as did his detailed, if confused, observations of the Least Seed Snipe. Beyond these few exceptions, though, Darwin's descriptions of birds were mostly brief and dull. Even obviously elegant birds, including raptors such as the Cinereous Harrier, which he watched in the Falklands, were mentioned offhandedly. "Falco —," he scribbled, "the only ones common in the Falklands." And then suddenly, or at least it seems entirely sudden, there was in Darwin's life and notes an explosion of both birds and ornithological exposition.

It is not merely that there were more birds on Darwin's lists but that Darwin was actually watching them with a much more persistent attention, and reporting back what he saw with increased depth, care, and competence. In one of his first few entries at Maldonado, Darwin described a bird he called *Muscicapa*. It had "habits like the common English flycatcher, but does not so generally return to the very same twig. . . . Beak, eyelid & iris beautiful primrose yellow. This bird is common over the Pampas, even so far as Mendoza at the foot of the Cordillera; it has not however crossed that barrier into Chile." This is the Spectacled Tyrant, a small black bird with a startling fleshy yellow ring around the eye. The next bird on his list was some sort of finch, with a "loud shrill cry; flight clumsy, as if tail was disjointed: base of bill dusky orange." (I am constantly tempted to see descriptive poetry in these colors but must remind myself that Darwin was working from a standardized nomenclature of color for naturalists, holding his dead specimens up to a book filled with squares of color, something like a giant paint chart, with codified names. So colors such as "primrose yellow" and "dusky orange," though lovely, are usually not Darwin's own.) After a thoughtfully described kingfisher came the Shiny Cowbird, which Darwin correctly placed in the Icterid family of orioles, blackbirds, and cowbirds. This one was glossy purple and black, not brown-headed like the common North American species. Darwin "heard many of them attempting to sing or hiss for I do not know what to call it. — The noise was very peculiar resembling bubbles of air from a small orifice passing through water. . . . I at first thought it came from Frogs."

From now on, this is what Darwin's ornithology looked like. He watched for distinctive behavior that might distinguish similar species. He carefully recorded geographic ranges. He paid attention to avian vocalizations long before it became popular to do so. He appreciated subtleties in morphology of both plumage and "bare parts," as ornithologists call them — bills, eye rings, and feet. What impresses me most is his ability to ask questions, to fix his gaze on a detail, even a small one, and to fit it into a theory, making it large. His conjecturing over the Shiny Cowbird was simple, but just the beginning of a pattern that would expand and deepen in the next three years. "In this same flock, there are commonly brown specimens," he wrote. "Are those one year old birds," he puzzled, "as amongst the Sturnus vulgaris? or females?" *Sturnus vulgaris* is the European Starling, the introduced starling of our own city sidewalks, which in Darwin's time was safely ensconced in Europe, where it should be. Male and female starlings look alike, as do many cowbird species, but the Shiny Cowbird is sexually dimorphic — males and females have different plumage characteristics. So the brown birds Darwin saw could have been either females or first-year birds, as he guessed.

I have pored over Darwin's letters from this time, as well as his field notebooks, his diary, and his specimen notes, again and again, looking for clues to explain his sudden ornithological interest. Surely it can't have just happened; surely there is some gradual metamorphosis, not simply an abrupt burgeon-

ing of recorded observation. I find myself somewhat dumbfounded by it, wondering what transpired between Charles
and the birds of Maldonado. But as I allow the clues within
his writings to sink in, it becomes clear that a number of
things did happen at Maldonado, and in the time leading up
to it, that allowed Darwin's latent flair for ornithology to surface.

For one thing, the birds there were easier to see. The landscape was a monotonous scrubby grassland, nothing at all like
the shady, secretive forests of Brazil. "The country continues
very similar," Darwin wrote to his youngest sister, Catherine,
"so that one dreadfully misses the georgeous views of Brazil."
Visual boredom might have prompted him to pick out birds
to keep his mind active, and in any case there they were — on
display, unhidden, ready to be seen.

In this vast expanse of active birds, Darwin began to take
pleasure in the way his collection was shaping up, and he enlisted help to keep the work moving. "My collection of the
birds & quadrupeds of this place is becoming very perfect," he
wrote delightedly in the diary. "A few Reales has enlisted all
the boys in the town in my service; & few days pass, in which
they do not bring me some curious creature." He was Sherlock
Holmes dispatching the Baker Street Boys, who returned with
the scrappy, wild knowledge that only loose, unsupervised children possess on their home turf. As if this weren't enough,
Darwin took on an assistant, a teenage boy who began the voyage as "Fiddler and Boy to the Poop Cabin." Darwin taught the
boy, Syms Covington, to shoot and prepare bird skins, freeing
himself from some of the dreary slog of collecting.

Though he had been making use of Covington's talents
with FitzRoy's blessing, Darwin wanted to remove any ambiguity from the relationship. In the same letter to Catherine,

Darwin inserted a hilarious business aside to his father: "Having a servant of my own would be a really great addition to my comfort." (Ha. Mine, too.) Darwin presented a detailed accounting of his expenses, explaining how he had been uncommonly frugal, spending only two hundred pounds per annum, and how he could employ Covington for a mere sixty pounds per annum. And since Charles surely couldn't be expected to wait around for an answer by sea mail, he had "come to the conclusion you would allow me this expence."

Without further ado, he dipped into his father's purse and secured the invaluable services of Syms Covington, with whom he developed a relationship that would continue several years after Darwin returned to London, until 1839, when Covington emigrated to Australia. In recent years, Covington himself has become a bit of a cult figure, with his own website and fan club of sorts. But at the time, he was just a boy, and doubtless happy to be working for a man more courteous than the bossy sailors, and to have some work that, although it involved a good bit of bird intestines, had nothing to do with deck scrubbing. His contribution to Darwin's work was enormous.

In a way, the Baker Street Boys and Syms Covington signal the new level of commitment exhibited by Darwin at this stage in the voyage. Where, at the beginning of the journey, he quietly worried over his ability to be entrusted with such an opportunity, he now boldly asserted his intent to do good, full work.

In the same letter to Catherine he confided sweetly, and with his disarming, Charlotte Bronte–like self-effacement:

I have worked very hard (at least for me) — at Nat. History, and have collected many animals, & observed many geological pheonomena; & I think it would be a pity, having gone so far, not

to go on and do all in my power in this my favourite pursuit; &
which I am sure will remain so for the rest of my life.

Almost hidden in this long letter — an expansive statement
of intent, a shy but solid recognition that he has done good
work — is Darwin's awareness that he stands on the threshold
of an incredible moment of opportunity and that what he ac-
complishes here will touch the whole of his future life. It was
his strongest statement to date of his sense of himself as a serious naturalist, very different from the excited dipping into
naturalist's clothes that characterized his first months on the
Beagle.

Part of his increased depth in avian observation came
from practice. His skills were increasing through intensive experience. He still loved his books, but it was his hands that
had begun to shape his fresh understanding. His knowledge
was becoming visceral, it was becoming unique, and it was becoming entirely his own. It was expanding in relation to creatures, and in relation to landscape. Perhaps Darwin had felt
little qualified to make detailed ornithological observations at
an earlier point, but now he was becoming sure of himself,
sure that he was seeing with a measure of insight, with a naturalist's vision that, though perhaps not entirely accurate or all-
knowing, was at least worthy of record. I almost envision this
emboldened young naturalist standing on a rock with his
arms spread wide. This was his first glimpse of an integrated
natural vision — all things necessary, all worthy of attention.

Within this new and wider line of sight, birds that he had
never seen kept tripping along his path. The birds themselves
were fabulous, insisting upon the significance of their presence. Finally Darwin was able to take them in. His statement
in the *Ornithological Notes* was simple and pure. "In this undu-

lating open grassy country, birds are very numerous," he wrote. "It is impossible not to be struck with the great beauty of the greater number of the birds; the most prevailing tint is yellow, & it is worth noting, that the same colour is strikingly characteristic of the Flora." The birds are many, they are lovely, and they are yellow like the flowers. And they sing, too, though "as songsters, the whole are miserably deficient" when compared with the English birds. In these lines, Darwin entered the naturalist's best place, attuned to the many dimensions of creature lives, allowing the birds themselves to evoke a sense of continuity between observer and landscape through layers of voice and feather.

Darwin began enjoying many sorts of birds here, several of them large and flashy, such as owls, skimmers, and rheas. But much of Darwin's best ornithology, and some of the loveliest moments in the development of his naturalist's eye, emerged from his persistent observation of the South American passerines. We often think of passerines as songbirds, though *perching birds* is a more accurate lay term. The passerine order is large, harboring extremely diverse families of birds. Tiny birds come readily to mind, such as the warblers and wrens, but in South America the cotingas and oropendalas are crow-sized passerines, and in North America the Common Raven, larger than many hawks, is the largest passerine species. The passerines have radiated so extensively that they are found in every terrestrial habitat on earth, excepting extreme Antarctica. All have similar feet, with three toes facing forward, and one pointing permanently to the rear for sturdy perching.

Within the passerine order, ornithologists distinguish between the oscine and suboscine passerines. The distinction wasn't noted until 1847, so Darwin wouldn't have had the framework for thinking of the birds as divided in this way, but he sensed something about it when he complained of the lack of accomplished birdsong in the tropics. All passerine species have complex syringeal, or vocal organ, muscles and are capable of producing vocalizations that are far more intricate than other sorts of birds (hummingbirds, storks, hawks — any others, in fact). But among the passerines, the suboscine species have a much simpler vocal apparatus than the oscines. Many of them are exceedingly vociferous, with a range of vocalizations, but only some of the oscine species produce the melodious vocalizations that we finicky humans deem worthy of being called song.

Of the more than thirty-seven hundred species of birds of all kinds in the neotropics, nearly one-fourth are suboscine passerines. Only a handful of suboscine species — a few flycatchers — are found in North America and Europe. They are considered to be more primitive than the more recently evolved oscine groups, and in addition to having a simpler syrinx, most suboscines exhibit a simpler humerus in the wing, simpler scapula in the shoulder, and even simpler sperm. Although the anatomical differences that distinguish the suboscines from the oscines are nuanced and in fact not visible in the field, there is something else — something not particularly tangible but nevertheless true — that draws the curious naturalist to these birds, something besides the fact that we may not have seen them before. While the oscine passerines are for the most part extremely similar in terms of their bodily structure, the suboscines are subtly less so. The suboscine passerines seem the tiniest bit "goofier" — a bit long-legged, or tall-headed, or large-eyed, or barely odd in some other way. There is something about these birds that makes us want nothing more than to befriend them, to follow them about all day. Darwin was certainly tugged along by the strange little birds, intuitively sensing their uniqueness and reveling, I like to think, in their newness.

Early on, Darwin was captivated by a group he called the *Certhia*, because of their resemblance to the small treecreepers in this family that he had seen back home. The two species of European treecreepers are extremely similar to the single North American *Certhia*, the Brown Creeper. They are small, hunched, white-bellied birds with slightly curved bills and stiff pointed tails, which they use in much the same way that woodpeckers do, as props for skirting tree trunks in search of insects hidden beneath folds of bark. The birds Darwin was watching were not *Certhia* at all but a suboscine group in the *Furnariid*, or Ovenbird, family called the spinetails.

The many spinetail species are all quite similar and difficult to make out. Modern taxonomists still argue over definitive names and relationships. Many species have a reddish crown, and some are reddish on their wings as well. They can be streaked or smoothly brown. Nearly all have long tails, with the middle feathers being particularly long and curving into an extended, sharp-looking point. Surely it was these wild tails, along with the general coloration, that made Darwin think of the *Certhia*, though, other than being two kinds of passerine birds, the *Certhia* and the spinetails are not remotely related. The spinetails, in fact, often have soft and even lax tail feathers. Since they perch in reedy haunts rather than shimmying on tree trunks, stiff, supportive tails are unnecessary for them.

Darwin became fascinated by birds' tails and closely observed their many uses. The Scissor-tailed Flycatcher, with its impossible streaming feathers, was a particularly wonderful example. "A forked tail is evidently of great utility in turning short," he wrote in the sort of note that would eventually influence his thinking in the *Origin of Species.* "We see it in the Frigate bird; the swallow, the tern, & Rhyncops" (or Black Skimmer). He studied hummingbirds, recording the ways that "the expansion of the tail be-

CURVE-BILLED REEDHAUNTER

tween each flap appears both to steady & support the bird."

Darwin was actually a bit unnerved by the spinetails' flimsy tails. In an oddly worded statement, he commented on one of his favorite spinetail specimens, one we now call the Straight-billed Reedhaunter, noting that the tail was remarkable for its "looseness of attachment." Regarding another spinetail skin, Darwin noted regretfully, "This specimen is tailless. . . . These feathers appear singularly liable to fall out." He managed to obtain another specimen of the same bird and preserved it in a jar of spirits to make sure he would bring at least one tail home. He also collected bits of tail from other hapless spinetails here and there, hoping, eventually, to glue them onto the poor tailless specimen (emobodying in such efforts the scrappiness that defines field ornithology at its best).

Darwin slowly began to see beyond the surface, to understand that appearances can be deceiving, that the color of a bird's body and the shape of its tail may have nothing at all to do with its relationship to other birds. "Certainly they do not resemble in their habits the true Certhias," he mused, and noting their strong legs, quite thick in proportion to their slender bodies, he guessed, tentatively, their true status: "Are they not remotely connected with the Furnarii?"

Since the several species of spinetail observed and collected by Darwin were very similar and their taxonomy was little understood, John Gould (who sorted Darwin's birds for the *Zoology*) had trouble making them out, and I have trouble making out Gould's names in relationship to modern ones. But it is clear at least that at one point Darwin held in his hands both the Curve-billed and the Straight-billed Reed-haunters. The birds are very much alike, but the Straight-billed has a bill that is longer, somewhat more slender, and sometimes straighter. (The degree of length, breadth, and curvature varies among individuals of both species.) Darwin was puzzled with good reason, holding a Curve-billed and thinking of a Straight. "I scarcely believe it to be a different species," he wrote, "more especially as I found one specimen, which was intermediate in character between them both." And regarding another specimen: "Only differs in shape of bill upper mandible in the latter is longer & the symphysis of the lower one is of a different shape in the two specimens." Then the wonderful question, so significant in the shaping of his life's thought (though Darwin had no overt sense of this at the time): "Are they varieties or species?"

What makes a species? A very young Darwin arrived at this splendid question in his pondering of bills and tail lengths and feathers. In these notes on the reedhaunters we can almost watch Darwin's mind as it turns from observed minutiae to broad questions, the theory and the details wondrously twined. Modern birders so often fail to get much further than a bird's proper identification. If we can manage that, we give ourselves a nice pat on the back and move on to the next bird. Here, Darwin could not properly name the birds, yet he observed each one with an intensity that I hardly recognize, so rare is such watching in modern circles.

It might seem an obvious sort of thing to say, but there is a vast difference between the nature of a decent answer and that of a truly good question. An answer's origin is not difficult to pinpoint, rising, at least in part, from within the question that called it forth. But an eloquent question is much harder to come by, and far more surprising. Given Darwin's limited background in ornithology, I am continually impressed, and sometimes a little jealous, that he somehow asks so many beautiful questions — questions that modern ornithologists still ponder and consider significant. In spite of what our teachers told us, there *are* stupid questions, but a good question, a beautiful and intelligent question, is always startling and treasurable. The best questions arise from a different sphere, from the unknown place that music or poetry comes from, the realm of creativity and curiosity and clear blue ether.

Responding to the questions conjured up in his own young brain, Darwin noted that the Double-collared Seedeater and other ground-feeding birds swallowed stones to help grind seeds during digestion and that these same birds had particularly strong gizzards. He observed cowbird brood parasitism (the practice of laying an egg in another bird's nest), considered its effect on the parasitized birds, and compared this practice in tropical cowbirds with that of the cuckoos of England. He observed the progression of plumages in birds of different ages and sexes. He used a bird's habits to aid in the differentiation of similar species. He noted that vocalizations change throughout the year, and he considered the purpose of ventriloquism in forest birds. Eventually, and still without any inkling of how important such questions would eventually be for him, he became deeply interested in the distribution of species, the way a similar species "replaces" another in a different geographic location. He watched and recorded, for every bird he saw, its manner of walking, perching, flapping, and resting. All of this was rooted in the naturalist's stillness that Darwin had begun to cultivate. It had to be; stillness is the only ground upon which such confidences are offered.

In Chiloe, Darwin studied another group of Furnariids, the tapacolos. One, the Chucao Tapacolo, is a particularly shy forest species that is very difficult to see. Modern ornithologists rarely lay eyes on it without relying on a recording of the tapacalo's voice to draw the bird out with its natural curiosity or defensiveness. But Darwin was developing a lighter touch. Chucao Tapacolos are small, and to observe them closely, Darwin made himself small, and quiet, and patient. "This bird frequents the most gloomy & retired spots in the humid forests," he wrote, and "at some times, although its cry may be heard, it cannot with the greatest attention be seen; but generally by standing motionless, in the wood, it will approach within a few feet, in the most familiar manner."

Such familiarity runs both ways, and it was becoming the center of Darwin's natural insight. Here Darwin spoke with a simple warmth toward his subject; he had entered into a quiet intimacy that allowed him ease and facility beyond the cool objectivity of pure science. Here, in patience, in stillness, the birds show themselves and tell their secrets. Their stories are not shaken out of them beneath a microscope but revealed, animal to animal, with a kind of earthen *familiarity*, on the forest soil.

Darwin's watching, as it matured, maintained an air of bright expectancy — a sense that, were he to watch even the smallest creature for years on end, it would still possess secrets beyond the reach of his knowledge. Such watching centers on a recognition of unplumbable depth, on a belief that all living beings are rich beyond measure. And so, in reading Darwin's notes, I find instruction in the art of field observation — and here I would include everything from scientific research to recreational birding to lunching in the proximity of backyard crows — hidden between the lines. Come, these notes say, be expectant, do not be dull, but bring the lost fullness of your intelligence to this endeavor, as you come quietly into the presence of wild things.

CHAPTER FIVE

Ostrich Soup

The ostrich, I believe the Cock, emits a singular deep-toned hissing note, which cannot be described. When I first heard it, standing in the midst of some sand hillocks, I thought it came from some wild beast.

— *Ornithological Notes,* BAHIA BLANCA, 1832

Several years ago, I had the rare opportunity to study seabirds on tiny Tern Island, one in a line of little projections of guano and coral that make up the French Frigate Sholls atoll in the Northwest Hawaiian Islands. At the time, Midway Atoll in the northwest chain had not yet been opened up to controlled ecotourism as it is today, so, along with four other Fish and Wildlife volunteer seabirders and a refuge manager (who was also one of my best friends), I was one of just six humans within five hundred oceanic miles of Honolulu. The island, a constantly shifting forty acres, is blanketed with seabirds, tens of thousands of them. There is a narrow gravel runway through the island's center, providing the only safe place to walk, away from monk seal pups, shy green sea turtles, and thousands of seabird chicks and eggs. As we walked the island from end to end, black-and-white Sooty Terns flew a few feet over our heads, crying our presence before us, facing us and flying, impossibly, backward. They landed on our heads, then wiggled their small, neat, triangular black feet under our hair, found an easy grasp on our scalps, and rode along.

Free from money and from anything on which to spend money — cars, films, restaurants, normal social outings of any sort — we became shockingly basic in our human interactions. We had work to do, but we had much more time than we had work. Sometimes we talked a lot, but oftentimes we didn't talk at all. We wore few clothes. We sunned ourselves and swam. We became a small troupe of lowland mountain gorillas, picking flies out of one another's hair.

I don't know what I was thinking of one particular morning, checking the eggs in the colony of Brown Noddies whose population we were monitoring, but certainly my mind was elsewhere. Perhaps I was pondering something that actually had to do with the terns themselves, like how distractingly cute and fuzzy their chicks were, or how unfortunate it seemed that female frigatebirds kept flying in, plucking the new chicks up with their sharp, curled beaks, tossing them into the air, and then deftly catching and swallowing them whole and squirming. (It wasn't a case of simple ecological balance — we believed there was a disproportionate number of frigatebirds; the noddy population was being decimated.) Perhaps I was scanning the sky for possible marauding frigates. Perhaps I was contemplating various troubles with my far-off boyfriend. Perhaps I was considering something hugely mundane like the progress of my tropical tan, for, in spite of my naturally pinkish tone, and though I tried to stay out of the sun, even my belly was turning a dark blackish brown, and this impressed me. In any case, absently tallying nests, eggs, and chicks, I stepped right into a noddy scrape, crushing the owners' one egg.

Fortunately for my own psychological well-being, the contents of the egg were oozy and breakfastlike, like nothing that resembled chick parts, yet. But the noddy was entirely distressed and walked myopic circles around the broken nest all day. When its mate returned, the bird that had been guarding the nest actually crouched and put its head down. I could attempt to mount a slender defense, noting that tern eggs are creamy gray and dark-spotted, designed to look like part of their pebbly substrate. They blend in remarkably well, surely I couldn't be expected to spot every one. But of course, that is exactly what ought to be expected of a tramping human in a tern colony. Not only was it my job in the everyday sense, what I had been hired to do, it was also my responsibility as a naturalist, one treading, supposedly lightly and in a spirit of learning, in a wild place. In truth, I just wasn't paying attention. The noddy and its mate laid a new egg within a week, and I, years later, still feel humbled, and careful, in light of the experience.

There is a tendency toward a deepening humility in the practice of the naturalist's faith. I use the word *tendency* because I feel that the more time we spend in simple observation of the natural world, the more, over time, over a lifetime, we lean toward an earthen humility. But this tendency is a rocky one, strewn with obstacles.

✳   ✳   ✳

*I*n July of 1832, after Darwin's encounter with the Brazilian forests but still several months before he began discovering the Maldonado passerines, the *Beagle* sailed for Monte Video, a port town on the coast of Uruguay. Darwin was reluctant to leave the Brazilian forests that sent his young naturalist's heart into simultaneous raptures and repose. He would have been more reluctant still had he known what was coming. The ship's landing was a little fifty-foot hill called the Mount — the most elevated place in the immediate region, and the landmark that gave Monte Video its name. Darwin surveyed the landscape from the top of the Mount and pronounced the view from the so-called summit to be "one of the most uninteresting I ever beheld." There was not a tree, not a dwelling, and seemingly not even a wild animal to break the monotony of the scene. "An undulating green plain and large herds of cattle has not even the charm of novelty," he wrote. Making matters worse, and to Darwin's "great grief," the region was marked by acute political unrest that made explorations of the countryside imprudent. Darwin was stranded for a time in the town of Monte Video with its "irregular and filthy" streets, broken only by a fruitless collecting trip with Captain FitzRoy to the nearby and aptly named Rat Island. Darwin, normally quite even-tempered, was crabby.

Finally, after nearly two weeks, a calm in both weather and politics permitted an amble in the countryside. Toting guns for sport hunting, Darwin and three of the ship's officers set forth, hoping to return with a brace of pheasants for a fresh meal. "And if our sport was not very good the exercise was most delightful," Darwin wrote in his diary. The group was also hoping to view a flock of what they called "ostriches." Though Darwin was, by now, the default naturalist on the ship, a refined interest in natural history was not uncommon among the officers of this early Victorian era, and the sight of something as spectacular as an ostrich was not to be missed. It was the first time they were in a place that ostriches were known to roam. No flock was to be found that day, but they did see a single bird in the distance, and Darwin confessed that if it were not for his more sharp-eyed companions, he would have mistaken the bird for a very tall deer. Even at this range, Darwin was impressed by his first ostrich sighting, and he thought the bird looked "more like a large hawk skimming over the ground. — the rapidity of its movements were astonishing."

In fact, Darwin was not seeing an ostrich. The bird in the distance was a Greater Rhea, a member of what we now call the Rheidae family, endemic to the neotropical region. Though the phylogeny of these birds remains controversial, it is likely that the rheas and ostriches descended from a common ancestor on the ancient southern landmass of Gondwanaland. When this continent broke apart into Africa, South America, Australia, and Antarctica at the end of the Cretaceous period, the family of large, flightless birds dispersed across the southern hemisphere, refined through evolutionary history into the species that Darwin saw, the ones we see today. Some ornithologists continue to argue that the rheas and ostriches are not related, that they developed similar bodily characteristics via convergent evolution. But studies of the species' morphology, genetics, and biochemistry strongly suggest a common origin for the two groups.

Rheas have more feathering on their heads, necks, and thighs than ostriches; and their thick, strong legs and feet are tipped by a different number of toes. Most birds have four toes, one too many for the rapid running central to the rhea's existence. Rheas have only three toes, and in the ostrich this number is further reduced to just two, calling to mind the early horses and the toed ungulates that populate similar habitats in these regions. Still, the birds are very much alike; it was commonsensical rather than ignorant for Darwin to refer to the South American rheas as "ostriches."

When, in early September, the *Beagle* sailed for Bahia Blanca and remained there a month, Darwin was for the first time in territory heavily populated by rheas. At the time of the *Beagle*'s docking, Bahia had been settled for only six years, and the beautiful, sharply carved bay had not yet been mapped. The town had been built up as a fortress against "Indians," Darwin recorded in his diary, and it served as a connection between Buenas Ayres and Rio Negro. The ship and crew were scrutinized thoroughly by the gauchos — a peculiar group of cavalry-cum-herdsmen typically of mixed native and European descent that peopled the South American pampas. Once the shock of seeing a vessel as large as the *Beagle* pulling into the harbor on their quiet shores had worn off, it was Darwin himself who became the center of the gauchos' interest and inquiry. He was not dressed in shipmen's weeds like all the others, he had no official title besides *un naturalista*, a term entirely unknown in Bahia, with a rough and unsettling translation: "a man who knows everything." Darwin was covered with vials and notebooks and beetles in bottles and all manner of odd, purposeless things. Eventually it became clear enough to the gauchos that Darwin was, as far as they were concerned, as harmless as he was useless.

For Darwin's part, the gauchos were some of the most wild and impressive beings he had ever beheld. They formed a "most savage picturesque group," he wrote. "I should have fancied myself in the middle of Turkey by their dresses." The gauchos were adorned with bright shawl "petticoats," fringed drawers, enormous spurs, and "Ponchos" — an unfamiliar word to Darwin, one that he carefully defined in his diary. Perhaps the most singular element of their costume was their boots, fashioned from the hide of the hock joints of a horse's hind legs, which formed oddly bent tubes. "This they put on fresh" — Darwin pondered the strangeness of it — "and thus drying on their legs is never again removed." The men themselves he found even more remarkable than their dress — muscular, with untamed expressions that recalled, to the rather sheltered young Englishman, "wild beasts."

In spite of the mixed impressions and reservations that ran both ways, the gauchos were civil to Charles, and regarding his newfound interest in "ostriches," they would become indispensable. More than Monte Video, Bahia was serious rhea country. "The ground," he wrote, "is tracked in every direction by ostriches."

Darwin gathered lucid impressions of the landscape and, in his pocket notebooks, twined them everywhere with rheas. "Ostriches tame, made sail." "Many ostriches, flocks from 20–30, beautiful on brown of a hill." Very pretty. The juxtapositions are sometimes odd: ". . . met on this road Indians from this place supposed to have murdered the postmaster. . . . Division of army to follow the track, if guilty to massacre them. Ostriches, males, certainly sit on eggs, easily distinguished, stray eggs first laid." The rheas gained an easy foothold in Darwin's imagination, and he began to observe, ponder, study, and occasionally eat them in earnest.

When a group of ten gauchos offered to lend Charles a horse and take him out hunting, he jumped at the chance and was so impressed with the fine ostrich "chace" he witnessed that he recorded it everywhere he could think of — in his diary, in his pocket notebook, in his specimen notes, and in his letters home. The gauchos spread out on horseback, several hundred yards between them, and encircled a group of rheas, slowly closing in on the birds, confounding their rather small brains. They employed a tool called a bola, made up of stones secured to the ends of leather ropes. These were hurled with a practiced spin at the rheas' legs; the long-legged birds would become hopelessly tangled and would fall to the ground, where they could be easily dispatched with a swift knock on the head.

After the hunt, Darwin was jovial around the campfire that the gauchos built at midday, cooking a fine meal of armadillo meat and ostrich dumplings. He lamented the fact that his English colleagues couldn't see him dining on such unusual fare in such an astonishing place and with such robust company. He ate with gusto.

Darwin does not mention culinary techniques in his notes, but it is likely that he was introduced to the traditional means of cooking both rheas and their eggs, a method still employed today. A large bird like a rhea is difficult to cook all of a piece over a fire — the outside layer of the bird will roast but little else. Instead, large stones are heated in the coals and then inserted into the rhea in place of the entrails to cook it from the inside out. Eggs are prepared directly in their thick shells. A wide hole is cut out of the top of the egg to allow for expansion and mixing with a stick or spoon. The whole thing is placed at the edge of the fire, and the contents are stirred occasionally until the egg is ready to be eaten.

Greater Rheas are impressively large, standing four feet from toe to forehead. They are highly adapted to running on the South American *campos*, with their strange, thick, three-toed limbs, and can hit speeds of almost forty miles per hour. Darwin loved their fluffy, loose feathering and the way they would "make all sail," a powerful description that he employed over and over to evoke the image of the rheas running with wings outstretched, a maneuver that helps them to maintain balance and control their speed.

The morphological features of the Greater Rhea were described thoroughly and accurately in Darwin's notes with a few curious exceptions. One of these is the rhea's lack of tail feathers. When ornithologists use the term *flight feathers*, they are referring to both the wing and tail feathers. In flighted birds, tail feathers are an important part of the flight apparatus, useful for steering and providing advanced aerodynamics. Rheas do not fly, and it seems evolution has trimmed the energetic task of producing feathered tails. Wing feathers are useful to the rheas in other tasks — not just for balance but for self-defense, too. Each wing boasts a strong claw that can be employed as a serious weapon, another characteristic that Darwin does not mention, though in adult birds it is peculiar to rheas. Strangely, he also does not mention the fact that the male rhea has a penis, which is often visible. Males of most bird species simply have a cloacal protuberance, an area on the lower abdomen enlarged throughout the breeding season for the transmission of sperm during coitus. This curious omission of Darwin's cannot be explained away by early Victorian prudery, which did not extend to the sciences. All animal parts were typically described with smooth objectivity.

The fact may be that Darwin did not observe the rheas when either the claw or the penis would have been in high use. The breeding season for rheas is August through January, with the most animated aspects of this process occurring in August and early September, including the gathering of mates, the actual mating, and the early, rigorous defense of the nest. Darwin began rhea-watching in Bahia Blanca in September, and many of the questions he ponders in his notebooks suggest that he just missed this crucial time.

The gauchos provided Darwin with a wealth of information on rhea nesting habits. The round nests, simple mounds on the ground, are filled to brimming with large, pale eggs. The rhea relies on its finely camouflaged feathering — the exact color of the *campos* plains — for cover. "The cock when on the nest lies very close," Darwin wrote. "I have myself almost ridden over one." The first nest Darwin saw contained twenty-seven eggs, and his guides assured him that this was not unusual, that they themselves had seen sixty eggs in one nest, and others had seen still more.

Darwin came to readily accept the natural history information gleaned from the gauchos, and he was fond of noting that "Indians and such people are excellent practical naturalists." He was impressed that the inhabitants of the countryside could readily distinguish the cock and hen rheas from a great distance. "The former is said to be larger and darker coloured, & its head bigger," Darwin wrote with the consternation of one who never quite learned to make out the difference for himself. In spite of his confidence in the gauchos' information, it took several insistent observers to convince Darwin of the fact that these oversized clutches were comprised of the eggs of many different females laying in the one nest and — just as intriguing — that the eggs were incubated by the male rheas, who not only hatched the eggs but also raised the young without female involvement. "The Gauchos unanimously affirm, that the male bird alone hatches the eggs & for some time afterwards accompanies the young," Darwin was pleased to record in the long rhea section of his *Ornithological Notes*. "I conceive there is not the slightest doubt on the subject."

He was still confused by the number of loose eggs left scattered about the plains, often quite close to a viable nest. "In one day's hunting, the third part were found in this state," he wrote. "It seems odd that so many should be wasted." The gauchos call these "walchos," Darwin jotted in his pocket notebook, but a fluent speaker eventually looked over his shoulder and corrected his spelling. *Huachos.* The intelligent, far-flung speculation that hallmarks the younger Darwin began to emerge: "It is evident, that there must at first be some degree of association between at least two females; else all the eggs would remain scattered over the wide plain at distances far too great, to allow of the male collecting them into one nest." The female rheas get together and lay their eggs in one spot — it's an interesting hypothesis, and it's almost right. The truth is even more colorful, and it's a pity Darwin didn't arrive just a few weeks earlier. He might have witnessed the rhea's breeding display, and he would have lapped it up.

Early in the breeding season, females do wander off in small groups, much as Darwin suggests, while the males become territorial, competing with one another in short-lived fights for the best nesting areas. They run around one another in quick circles, make terrible hissing sounds, and sometimes bite and kick. As is typical in avian society, this aggression is largely symbolic, and serious blood is rarely drawn. But there is enough activity to establish dominance, and the victorious male runs all other cocks out of his domain. He prepares a nest, a depression about a yard wide and a foot deep, often near some scrubby shrubs to aid in camouflage. Frequently, the rhea uses his strong legs to uproot the vegetation in a two-

or three-yard circle about the nest, presumably to isolate it in the event of a fire.

He then attempts to attract females into his territory by running at them with his wings outspread and fluttering, attempting to look (we now know from later Darwinian principles) like as strong a breeding specimen as possible. He assembles up to twelve females, and moves on to the day's highlight — performance of the true courtship display. The dance involves zigzag prancing around the group of hens while emitting a baritone "boom" voice reserved for this occasion. His wings are fluffy and lifted, his neck erect and inflated. For his finale, the rhea stands quite still, head lowered, neck curved in a lissome U. He gives his wings a last vigorous shake, then holds them still and outstretched with an ever-so-slight quiver, confident that no right-thinking rhea female could resist such a display.

After mating with the females (who actually do seem quite overcome), the male rhea leads them to his nest, where they each line up to lay an egg, one after another. The females wander off together and return every few days to lay another egg until the clutch is complete. The male commences incubation, while the females move on to mate with a different male, repeating the process at another nest.

Examining a female rhea that one of his shipmates brought home for dinner, Darwin counted fifty eggs of graduated size in her oviduct — a terrific number of eggs to be laid in the course of a single season. The rhea nesting arrangement guarantees that the eggs in one nest are laid at roughly the same time and can be hatched together. In spite of being unable to observe the actual breeding and nesting, Darwin speculated quite rightly: "If the hen were obliged to hatch her own eggs, before the last was laid, the first probably would have been addled; but if each laid a few eggs, at successive periods, in different nests ... then the eggs in one collection would be nearly the same age."

While setting the clutch, the cock is intensely aggressive. "It is asserted that occasionally at such times, they are fierce & even dangerous, that they have been known to attack a man on horseback, trying to kick & leap on him," Darwin documented in his *Ornithological Notes*. "My informer pointed out to me an old man, whom he has seen much terrified by one chasing him." Poor chap. Even the female rheas approaching to add eggs to the clutch are often attacked by the males (no one ever claimed rheas were brilliant) and forced to lay their eggs outside the nest. These are Darwin's *huachos*.

"Some have believed that the scattered eggs were deposited for the young birds to feed upon," Darwin noted, unaware that the male's aggression forced their scattered presence. He went on to say, "This can hardly be the case in America, because the Huachos, although often times found addled and putrid, are generally whole." In fact, Darwin's instinct to find some ecological role for the *huachos*, or a "place in the economy of nature," as he would have put it, was not misplaced. Though young rheas do not eat their would-have-been siblings, they do feast on the flies that the addled eggs attract, and these protein-rich insects form a substantial part of the chicks' early diet.

All of this was terribly absorbing to Darwin, and he was sure it would prove to be of interest to the natural science community back in London. Many of his observations were new, and all of them added depth to the little that was previously known about the South American rheas. But he was more than dimly aware that his research was not entirely original. Rheas were not a species new to science. The spectacular birds had been collected and their basic habits described by explorers going as far back as the 1750s. Emus, Australian birds that are related to rheas, were kept in the botanical gardens at Kew in England, and their similar nesting habits had been carefully described.

About this time Darwin became aware of naturalist Alcide d'Orbigny's French-funded six-year expedition to South America. D'Orbigny had scoured the continent, amassing an enormous collection of avian, mammalian, botanical, and geological specimens. He returned to France just prior to Darwin's arrival, making him the closest thing to a nemesis that Darwin ever had. And who, in all honesty, does not secretly desire a nemesis, at least for theatrical purposes? Having read d'Orbigny's report, and milking his own native flair for drama, Darwin reveled in a letter to Professor Henslow: "I experienced rather a debasing degree of vexation." The French explorer described many of Darwin's own observations in annoyingly exact detail. Darwin had had "some hard riding for nothing." But there was one rare avian treasure that d'Orbigny, in spite of much searching, never managed to find.

Darwin first heard of the bird in Rio Negro, where he spent July of 1833. He wrote in the *Ornithological Notes*, "I repeatedly heard the Gauchos talking of a very rare bird which they called the Avestruz Petise." The gauchos described the bird as being smaller than the common rhea, which was abundant in Northern Patagonia, but bearing a "very close general resemblance" to the larger bird. The color of the smaller rhea was darker, and "overo," or mottled. Its legs were shorter, Darwin was told, and feathered farther down. By now, Darwin did not doubt the information imparted by his native guides, and he was impressed that "the few inhabitants who have seen both kinds affirm they can distinguish them apart from a long distance."

The fact that this was likely to be the largest bird on earth that had still not officially been described in the scientific literature was not lost on Darwin. He was well aware that research into the Avestruz Petise, particularly if he could obtain a specimen, would reflect nicely on him back in London. But it seemed that in Rio Negro the best-known thing about the Petise was that it was an extremely rare bird. Darwin kept gathering facts, jotting notes in his little books, and searching for the Avestruz Petise without success.

There were a couple of artifacts in Darwin's collection that he regarded as genuine Petise parts. These were two large, dark feathers and a swath of hide he purchased from some native women in Port Desire late in 1833. Here again, Darwin relied with great trust on the naturalist capacities of the native people for the accuracy of his specimen labels. But for all his effort, he saw the living bird only once, and he wasn't sure enough of what he was seeing at the time to enjoy it fully. The instance is recorded in one of his little pocket notebooks. On April 23, 1834, Darwin "saw an ostrich about ⅔ size of common, and much darker coloured — exceedingly active and wild." Knowing that he was in Petise territory, and that these birds are more nervous and untamed than the common rhea, Darwin clearly believed himself to be in the presence of the rarer sort (and modern range maps for the birds confirm that in Rio Santa Cruz he would have seen no others). Aware that most European naturalists still recognized only one species of South American rhea, Darwin prudently refrained from overstating the case. The entry in his diary reads simply, "Ostriches are not uncommon here, but wild in the extreme." This was Darwin's first glimpse of a living Petise, but it was not the first one he'd seen, so to speak, in the flesh. That happened a few months earlier.

In an uncharacteristically cryptic entry in his diary, Darwin barely mentions the incident. It was the third of January 1834, and the *Beagle* had been anchored at Port Desire for nearly a month. "During these days I have had some very long and pleasant walks. — The Geology is interesting. I have obtained some new birds & animals." This last is a curious understatement, and we can only guess that in the moment, Darwin failed to see either the humor or the significance of the true story, which he later recorded in his *Ornithological Notes.* In these notes, the story becomes a virtual palimpsest, with deletions and additions coming and going, and even more mending after his return to London. In a final corrected version, this is what he wrote:

> When at Port Desire in Patagonia (Lat 48) Mr. Martens shot an ostrich; I looked at it forgetting, at the moment, in the most unaccountable manner, the whole subject of the Petise, & thought it was a two-third grown one of the common sort. — The bird was cooked & eaten. — & my memory returned. Fortunately the Head neck legs, one wing & many of the larger feathers had been preserved. From the fragments a surprisingly good specimen has been put together, and it is now exhibited at the Museum of the Zoological Society.

It is amusing to find, in this early telling, how objectively Darwin recorded the cooking and eating of his most sought-after avian specimen. In truth, it must have been an extremely colorful scene. We can picture a small group of the ship's officers serving up a delightful ostrich stew, pleased with the steaming tastiness of their meal until the feast is interrupted by their eccentric ship's naturalist, who is always doing odd things, and so it comes as little surprise, really, when he jumps up and yells, "Wait!" Darwin pulls the grisly bits of what was quite recently a complete Avestruz Petise out of the soup pot and the kitchen trash. One cannot help thinking that Darwin's estimation of the resulting specimen — "nearly perfect," he will write in the *Zoology* — must be a bit of an overstatement.

Once it was shipped home, in fact, the specimen had London's finest naturalists, including the well-known ornithologist and artist John Gould, fixing it up into respectability. Gaps in skin and feathers were filled in with matching feathers from other birds, including the more common rhea, and it really did look quite tolerable. Darwin enlisted Gould's help in officially describing the new rhea, and the two of them presented the bird to the Zoological Society of London in March of 1837, six months after Darwin's voyage had drawn to a close. The Society transactions document the event in two tightly worded pages, just following Professor Richard Owen's report on the olfactory sense of the Vulture. (Owen was at the time a supporter of the young Darwin's work and reputation. Later he became a rabid opponent of his eventual theories, often attacking him personally and publicly.) "Mr. Gould brought before the notice of the meeting, from the collection of Mr. Darwin, a new species of *Rhea* from Patagonia," the secretary recorded. "Mr. Darwin then read some notes upon the *Rhea Americana*, and upon the newly described species, but principally referring to the former." Darwin was pleased beyond measure with the next development. "Mr. Gould, in conclusion, adverted to the important accessions to science resulting from the exertions of Mr. Darwin, and to his liberal-

ity in presenting the Society with his
valuable Zoological Collection; to
commemorate which he proposed to
designate this interesting species by
the name of *Rhea Darwinii.*"

Unfortunately, the appellation
did not stick. Rules that grant official
precedence to names that are be-
stowed by the first scientist to describe a specimen would not be

LESSER RHEA

codified until 1841, but by 1837 that tradition was already
strongly in place. All concerned were dismayed to discover
that Darwin's naturalist nemesis — the French explorer Alcide d'Orbigny — had taken the liberty of naming the bird
himself. He called it *Rhea Pennata* (in modern times only the
generic name would be capitalized, but scientists in the nineteenth century typically capitalized the specific name as well),
pennata meaning "feathers" and referring to the unique feathered tarsus of the smaller rhea. There was a flurry of discussion, with names, feathers, and egos flying in all directions.
But in the end, the Londoners chose to ignore d'Orbigny's appellation, justified by this notice in their own *Magazine of Natural History:*

> We observe by a letter which lately appeared in one of the
> French Journals, that M. D'Orbigny claims the right of
> having first described the Rhea brought home by Mr. Dar-
> win from S. America, and which Mr. Gould named, a few
> months since, R. Darwinii. It appears that M. D'Orbigny
> gave it the specific appellation of R. Pennata, but in his
> letter he does not refer either to his published characters,
> or to the specimen which he examined.

With no dead bird to show for himself, d'Orbigny had no naming privilege. Though the London scientific community dismissed d'Orbigny's claim by way of its dubious foundation (and, doubtlessly, because of unabashed favortism toward one of their own), history was not as kind, at least in this instance, to Charles Darwin. In modern times, the common name of Darwin's Avestruz Petise is the Lesser Rhea. Its scientific name is *Pterocnemia pennata.* The generic name reflects an updated taxonomic understanding, and the specific name recognizes d'Orbigny's precedence.

The rheas are, in many ways, the lost story in Darwin's development of natural selection theory. Everyone knows about the Galápagos finches, but almost no one knows about Darwin's extensive involvement — both on the land and in the brain — with the South American rheas. And while the finches were never even mentioned in the *Origin of Species,* the two rheas became one of his flagship examples of the importance of geographical distribution in the study of species. He referred to them several times in the *Origin.*

His first comments were quite basic. When he was in South America, Darwin noticed that, though the rheas did not fly, they still used their wings extensively. This led to a graceful gathering of wing lore in the *Origin of Species.* Of course birds use wings to fly, Darwin mused, but penguins use their somewhat flattened appendages as flippers or as "front-legs" on land; loggerhead ducks use their heavy, broad wings for flapping on the waves; and rheas use their lacy wings as sails. Darwin stated clearly that these "grades of wing structure" should not be taken as intermediary steps on the road to avian flight. It was much more likely, he conjectured, that these birds descended from flighted ancestors, and grew into flightlessness over time (which turns out to be true). He used these varied wings and their functions to show instead "what diversified means of transition" between species are possible.

One conclusion Darwin drew about the rheas is enjoyably incorrect. The ostriches could not fly, he speculated, because "an enormous supply of food" would be necessary to give them enough force to get up in the air. Perhaps if the rheas ate more lizards or small birds for protein, they might gain some altitude. We now know that flight is more a matter of physics as it is a matter of food energy, and no earthly diet would support a rhea in flight.

Later in the *Origin* Darwin tied the two species of rhea directly to his theory of descent with variation. Darwin began his first notebook on the subject of natural selection shortly after returning to London, and it contains several jottings that link Darwin's observations of the rhea directly to his emerging transmutationist thoughts — and they *are* jottings: quick, vital thoughts flowing right from the brain to the pen in an effort to capture them quickly for future use. Darwin's granddaughter Lady Nora Barlow, who transcribed and carefully studied these notes, points to one passage in particular: "'When we see Avestruz two species. certainly different. not insensible change: — yet one is urged to look to common parent? Why should two of the most closely allied species occur in the same country?'" Here is an intimation of descent with variation in the two species with a "common parent," as well as a very early foundation for natural selection, the mechanism for evolutionary change that Darwin eventually described, but did not yet comprehend, in the question of "closely allied species . . . in the same country." With the rheas as his flagship example, Darwin will write in the *Origin*, "We see in these facts some deep organic bond, throughout space and time, over the same areas of land and water, independently of physical conditions. The naturalist must be dull who is not led to enquire what this bond is." And lest the naturalist remain unsatisfactorily dull, Darwin spells it out in the next line: "The bond is simply inheritance."

I absolutely love this winding story of Darwin and the rheas, but even more interesting to me than the Avestruz Petise specimen and its eventual name is the fact that, back in Port Desire, Darwin forgot about the smaller rhea in the first place, actually nibbling its giblets before his presence of mind managed to return. "Unaccountable," Darwin calls his own forgetting, and it is. After so much work and mental effort regarding the natural history of the smaller rhea, how could he have simply forgotten about the bird over ostrich stew? Was he regaling his table fellows with a fascinating tale from his week in the wilds? Was he preoccupied with some sort of scientific problem — something geological or paleontological? Was he mooning again about Fanny? Was he contemplating his tropical tan?

Darwin could laugh at himself as well as the next gentleman scientist, and this would become one of his most oft-recounted tales back in England. But it is more than a good story. In this "unaccountable" moment, Darwin was reminded of our capacity for forgetfulness, a capacity that can be shocking. Like Darwin, we tell stories of our own forgetful moments in the presence of friends, and we tell them lightly. "Can you believe what I did?" But in the moment after the telling and the little laugh, there is an unseen moment of solitude, a shadow of quiet, where the possibility of what our own forgetfulness could have meant, or might someday mean, rests in us. Typically, arrogance is considered to be the antonym

astonishing and sometimes mysterious nature as humans who live in culture but remain rooted in the soil, indigenous, at home in the lively wilderness of biological life. In humility we remember this condition, the earthly dimension of our peace, our truth, and our grace.

The practice of observing the natural world — of getting down on one's hands and knees before a tide pool, a lichen, a quail, a silent stone, learning from such wild things all one can about their place, their life, their needs, and doing this alone, or with a child, or with a lover, and doing this over and over again, over days and years — is humility's medium. In such moments, our vision is renewed, our sense of our proper place in the world is both strengthened and deepened. In a runaway social structure that gains its own strength by turning us away from our biological roots, we open, each in our own way, and turn back again. It is wonderfully counterrevolutionary to focus on the wild things that draw us, to know the birds, to learn them slowly, to study the native plants, one by one, to notice the insects that live on the undersides of leaves, to understand something of their tiny lives with no skull, and to do this, not in a rush, but sustainably, over time, over years, at the gentle but insistent pace of nature's own unfolding.

Darwin spent five years doing very little besides crawling around doing this, but for most of us such a proposition is impossible. However, in our own lives as homespun naturalists, the moments we do manage to spend becoming educated by our native places can wend their way into our daily lives, making it more and more difficult to see ourselves as individuals, self-sufficient and cordoned off somehow from our humus-y ground. We begin to see, rather, our lives as embodied, unseparate, inseparable, rushing forward with the whole of wild life.

I am constantly brought back to this understanding — that my own little life twines with the expanse of wildness — and feel it sometimes hanging all about me, like Darwin's rainforest lianas, in a lovely, necessary tangle. There is a complex line that circumscribes human interactions with the natural world — a line between remembering ourselves and forgetting ourselves that must be trod lightly and with discernment developed over time. I am more and more certain that this is what the naturalist's calling asks: to see how deeply it matters that we are not forgetful. We are charged with the simple yet strangely difficult task of watching where we step.

CHAPTER SIX

Come, Little Friend

I must now tell you what I think of him. . . . He is the most open, transparent man I ever saw, and every word expresses his real thoughts. He is . . . perfectly sweet-tempered and possesses some minor qualities that add particularly to one's happiness, such as not being fastidious, and being humane to animals.

— *Letter from Emma Wedgwood to an aunt,*
upon her engagement to Charles Darwin

The Great Kiskadee is a large bird in the Tyrant Fly-catcher family, a New World group. There is no comparable group in Europe, so Darwin would have had no internal reference for this species that he observed in various places throughout his journey. It easily caught his attention, being so tall and busy and brightly colored — a lemon-yellow breast, black-and-white-striped crown and cheeks, and a cedar-brown back. The bill is heavy, and the bird is smart, always active, always occupied, always up to something. Darwin thought its general habits very much like the "butcher birds," or shrikes, the only group of passerine species that actually prey on small mammals, impaling them on thorns for storage (something kiskadees don't do, though they do actively forage from air, trees, shrubs, or ground and are happy to eat insects, tadpoles, fish, and other hapless small things). Beyond this, Darwin often saw the kiskadee "hunting a field like a hawk, by hovering over one spot & then proceeding onward to another," though it did not "stoop so suddenly." Then, too, the kiskadee reminded him of a kingfisher. "Commonly it haunts the neighbour-head of water, & will in one place remain like a kingfisher stationary, it thus catches small fish which happen to the Margin." The kiskadee is dazzling and odd, a wild combination of behaviors and habits rolled into one yellow bird, and all of Darwin's associations make some sense. "The Spaniards say it is like the words 'Bien te veo' (I see you well). & accordingly have given this name to the Bird." *Bien te veo*, Darwin might have said to the kiskadee himself, and what better compliment could there be, really, from human to bird?

With its terrific behavioral range, the kiskadee made a good learning bird for Darwin. It was easy to watch, easy to accommodate to his new custom of recording not just what a bird looked like but also what it did, the habits of its life being equally as significant to the naturalist's task. From all of this Darwin gathered that the Great Kiskadee was a bird of remarkable "tameness" and that it was "cunning," with "odd manners." No wonder, he surmised, it was so readily kept by children.

Great Kiskadees are a bit unsettling to watch because, it seems, they so often watch you back. They peer deeply into your eyes with a slight turn of the head as if ready to ask, "How did you enjoy the film, Ms. Haupt?" Perhaps in French. It is easy to see why Darwin called them cunning. It is a bit like the pot calling the kettle for Darwin to notice anyone's "odd manners," but there you have it. One odd bird watching another.

As is typical of such turnings, there is no definitive moment to which we can point. But Darwin's writings suggest that somewhere around the time he sailed for Rio, two months into his voyage, his relationship to the animals he observed began to show a new facility. Where he had begun in a frenzy to establish himself within the predefined scope of a naturalist — as a scientist presiding over the nature about him — the actual experience of extended, close contact with wild things seems to have softened him. He began to approach creatures, whether he knew much about the particular being before him or not, with a kind of expectant familiarity. Just as he might have cordially shaken the hand of an acquaintance at Cambridge, a familiar gesture between gentlemen, Darwin greeted the creatures of the tropics crouched quietly on his knees, one animal to another.

From this vantage point, Darwin did not simply report the behaviors of a creature but went further — suggesting a connection between an animal's activities and its state of mind. In the *Ornithological Notes* he wrote that the Hornero was "fond" of dusting itself, just as certain hummingbirds were "fond" of retired, shady spots. A tinamou was "silly," and a Black Vulture waited "patiently" for its meal. This same vulture was not only tame and rather inactive, it was "cowardly." Certain species of spinetails were "industrious."

Such language came effortlessly to Darwin, and the

more time he spent in the presence of
wild things, the more frequently it
emerged in his writing and thought.
In Cambridge, his beetle collecting
had been zealous and even joyful but
marked by a kind of necessary dis-
connect between the collector and
the collected. The camaraderie there
was with other men, other collectors — Charles and his cousin

WHITE-THROATED
CARACARA

William Darwin Fox based their dear friendship on the shared
happiness of unearthing various rare species of *coleoptera*, and
then crowing about it. Here, in the near-constant company of
things utterly wild, and quite apart from the thin distance that
constant feedback between humans can impart, Darwin entered into what Nora Barlow called "sympathetic participation" in the lives of the creatures he observed.

In Darwin's time, anthropomorphism among scientists
was not considered a crime, as it became in the following
century (and largely remains today). Even so, Darwin's own
language regarding animal consciousness was unique —
unaffected, unforced, and original. He was not imitating
Humboldt here or other naturalists studied. From the time he
was quite young, it seems, there was an easy, innate tendency
in Darwin to recognize consciousness in nonhuman animals.
On the *Beagle* journey, Darwin's extended, direct experience
with creatures of all kinds allowed this tendency to flower.

I am often touched by watching how focused Darwin
could be on qualities that require a certain observational sub-
tlety. Standing gulls were "peaceable," petrels approaching the
ship were "tame & sociable, & silent," a certain thrush was
particularly "inquisitive." But Darwin himself was just as

eager to report "boldness" and "passion." The several species of caracara — sleek scavenger-raptors — provided particular fodder. While the "Carrancha," or Northern Caracara, was an "inactive, tame, cowardly bird" (cowardly, it seems, in spite of the fact that it "destroys young lambs by tearing the umbilical cord," and that it pursues the "Gallinazo," or Black Vulture, "till that bird is compelled to vomit up the carrion it had lately gorged"), the Striated Caracara was considered by Darwin to be extraordinarily "fearless." He defended his assessment with the story of a wounded cormorant, a bird somewhat larger and certainly heavier than a caracara, that was seized by a small group of the raptors and pounded on the head to "hasten its death." This doesn't sound so much braver than forcing vultures to vomit up their dead lambs, but Darwin was insistent on the point.

Apart from being brave, the Striated Caracaras were "mischievous and inquisitive." Darwin described their playful actions, picking trifles and treasures from the ground — a compass, a bola, anything that struck the caracaras' strange raptor minds — looking them over carefully, and sometimes making away with them. These birds were, moreover, "quarrelsome & very passionate, tearing up the grass with their bills in their rage." Darwin was a bit shocked at the caracara's crop, which would protrude dramatically after a carrion meal, "giving to the bird a disgusting appearance." Yet he couldn't look away. The caracara entries are among the longest and most detailed in the *Ornithological Notes*.

Various birds in Darwin's notes were praised for their peacefulness, inquisitiveness, intelligence, ingenuity, industry, and general friendliness. But there were also birds he simply found annoying. In La Plata he observed the Southern Lapwing, "called by the Spaniards Pteru-Pteru, in imitation of their cry," a bird closely related to the equally loud killdeer of North America. After making a few rudimentary observations regarding the behavior of the Pteru-Pterus, Darwin gleefully reported that they "appear to hate mankind & I am sure deserve to be hated, for their never ceasing unvaried, loud, grating screams." Not only were they loud, but they "pursue & fly round the head of anyone who invades their haunts." Darwin was further disgruntled by the fact that lapwings undermine the task of both the sportsman and the naturalist, by "telling every other bird of his approach."

Though I am particularly interested in Darwin's appreciation of avian consciousness, he was equally attentive to subtle personality quirks in a variety of species spanning the phylogenetic scale. He was reluctant to eat armadillos, they were "so quiet" (though not quite reluctant enough to *actually not* eat them). He was startled to observe the seals at Tres Montes, blanketing the coastal rocks in such numbers. "They appear to be of a loving disposition," he wrote in his diary, "and lie huddled together fast asleep like pigs." Darwin observed numerous bizcachas, large, white-cheeked South American rodents, noting that "late in the evening they come out to play." He strode up to them, and looked straight into their faces. "In the evening [they] are very tame, you may ride quite close, without disturbing the gravity with which sitting in the mouth of their holes they watch you."

For Darwin, the fact that animals are conscious beings was a given, not an assumption or a guess but a fact so basic, so obvious, that he did not bother to question it. In his life's work, he sought to elucidate the nature of both human and animal consciousness but never to defend the very idea. Based on commonsense observations, Darwin recognized that while animals are not little people, do not possess human minds or approach human language, they do have animal minds, a consciousness appropriate to their own particular being and place, and often express themselves in ways that, if we pay attention, we might recognize and understand.

Darwin himself rejoiced in this understanding, which deepened through the experience of his journey and developed even more fully in a lifetime of thought and work with nonhuman animals. An unabashed recognition of animal consciousness was not an eccentric little adjunct to his evolutionary work, as so many modern scientists, embarrassed, are eager to make it out. Such thinking, rather, is clearly in line with a developing evolutionary perspective where there is no break between human and animal lineages regarding either biology or consciousness. Darwin's sense of animal minds, and the expansive interpretations of animal behaviors that it allowed him, is one of the foundations of his evolutionary theory. Darwin would never have dreamed of teasing the two apart.

Nevertheless, many of the best Darwin scholars, and certainly most modern biological scientists, see Darwin's ready anthropomorphism as a charming aside to his serious work in evolutionary theory. Something to smile upon, indulgently, but certainly something we have outgrown in our scientific prowess. Randal Keynes has been a professor of physiology at Cambridge and an invertebrate specialist and is Nora Barlow's godson. I greatly admire his extensive work with Darwin's primary notes and make constant use of his beautifully edited versions of Darwin's diary and zoological notes. Keynes's remarks on this score are typical. Some of Darwin's observations, Keynes wrote in his recent introduction to Darwin's zoology notes from the *Beagle,* are "described in more anthropomorphic terms than would be acceptable today, but this does not detract from the liveliness of his accounts, or from his purposeful correlation of behavior with details of structure and environment." Not only does it "not detract," but, I find, very often Darwin's ready anthropomorphism forms the basis of his accounts' "liveliness."

The important point here regards a claim to the parameters of scientific relevance. The prevalent attitude among scientists today is that Darwin's attention to animal minds may be sweet and quaint, and it need not distract us from Darwin's *good* work, but thankfully we have progressed beyond such error. Biological science, structuring so much of its understanding of earthly life entirely within the strictures of evolutionary theory, is quite willing to recognize physical continuity along the phylogenetic scale but prefers to limit the official ability to "think," or even to possess consciousness, to humans alone.

From the start, most biological scientists are trained under a specific rubric that allows a certain range of intellectual speculation. Within this range, however, there are two things that are banned in scientific reporting. One is anecdote, the random, nonreplicable observations of animal behavior. The other is anthropomorphism, the attribution of human mind states to nonhuman creatures. Most young would-be scientists take their first college classes under a posted commandment that actually forbids such pitfalls. At my college, the sign read "Thou Shalt Not Anthropomorphize" and was penned in ornamental calligraphy upon faux parchment, as if it had been inscribed by a medieval monk taking dictation directly from God. Socially, culturally, and in the realm of good common sense, we have never really accepted the Cartesian dichotomy between the human and the nonhuman, where humans are the only thinking beings on earth and animals are more like mindless automatons, trapped senselessly in a world of stimulus and response. But this mind-set is still deeply ingrained in many scientific circles.

In recent years there have been signs of a thaw. Some ornithological researchers are beginning to admit "play" as a valid behavioral category, at least for species we consider more intelligent, such as ravens and crows. Early in the new millennium, MIT researchers published evidence that lab rats dream. We have Koko the gorilla, Alex the African gray parrot, and Rico the border collie teaching us the linguistic and logical capacities of their ilk.

Regarding birds in particular, a fascinating international gathering of scientists, calling themselves the Avian Brain Nomenclature Consortium, is working to move human understanding of avian brains and lives forward. Arguing that the way we name things has a powerful influence on both the way we think and the way we carry out experiments, the consortium brings to light new research into the evolution of the vertebrate brain to support their claim that the language used to describe the subdivisions of the avian cerebrum, and hence our understanding of avian consciousness, is in need of drastic revision. More than one hundred years ago, German scientist Ludwig Edinger, widely considered to be the founder of modern neuroanatomy, described a unified theory of brain evolution that underlay a nomenclature for describing the brains of all vertebrates. He was inspired by *The Origin of the Species*, and evolution as Darwin envisioned it. But Edinger distorted Darwin's ever-branching tree of life, overlaying it with a version of Aristotle's *scala naturae* in order to support his own view of evolution as progressive and unilinear, with animals rising in form and function, from fish, to amphibians, to reptiles, to birds, to mammals, to primates, and finally to humans. As vertebrate groups progress up this chain of being, brain size and complexity increase, culminating in the human cerebrum. According to Edinger, the avian cerebrum is made up almost entirely of basal ganglia, and therefore bird behavior is largely limited to instinct. The more malleable mammalian behaviors require a neo-cortex, which birds do not possess. Although heaps of research in recent decades suggest that this view is entirely incorrect, that the avian cerebrum functions in ways very similar to the mammalian cortex and that the cognitive abilities of many birds not only match but actually surpass that of many mammals, the language that modern neuroanatomists use to describe the avian brain remains unchanged. This primitive language, the consortium argues, keeps science locked in an outdated understanding of avian consciousness; they have proposed a new nomenclature for the subdivisions of the avian cerebrum, published in a well-respected journal, and, after a flurry of post-publication attention, are waiting impatiently for the neuroscientific world to meaningfully respond.

While certain very well-known thinkers, such as Tufts University philosopher Daniel Dennett, continue to insist that the scientific consideration of animal consciousness amounts to a "wild goose chase," others, such as Marc Hauser, an evolutionary psychologist at Harvard, insist that "the questions for the future, then, are not 'Do animals think?' but 'What precisely do they think about, and to what extent do their thoughts differ from our own?'" Donald Griffin, an ethologist of international repute and a kind of elder in the field of animal ethics, keeps working to make the obvious more obvious. "There are no neurons or synapses in the human brain," he continually reminds the unwilling, dragging them into the future Hauser speaks of, "that aren't also in animals." Still, a flip through any typical peer-reviewed biological journal normally will not bear mention of animals feeling pain, or pleasure and will give just a very occasional nod to "pain," or "pleasure" in the higher vertebrates. Animal consciousness tends to be relegated to the land of quotation marks.

In his wonderful book *The Unheeded Cry*, published by Oxford University Press in 1989, philosopher Bernie Rollin argues elegantly that animal consciousness was not banned from scientific consideration because of any new knowledge about animal thought processes or lack of them. On the contrary, the motivations did not issue from the imperatives of science, or of emprirical study, but from those of academic fashion. As quantitative physical sciences gained ascendancy, it was mathematical certainty, and a science that exercised control over its own outcomes, that gained preeminence. In an attempt to remain current, the "softer" fields of biology and psychology rushed to fit the quantitative requirements of the so-called hard sciences. A ready discussion of consciousness, and the anecdotes that so often support our everyday understanding of animal lives, had no place in this new world and was banished with remarkable efficiency.

I met Professor Bernie Rollin on the day I arrived in Colorado to begin my master's degree in the philosophy of science and environmental ethics. Rollin would be one of my advisors, so I thought I might as well introduce myself. I had read his first book, *Animal Rights and Human Morality*, in which he argued for a rights-based animal ethic grounded in a Kantian metaphysics. Then, as now, Rollin was rightly regarded as the "thinking person's animal ethicist," avoiding both the cold utility of animal ethicist Peter Singer and his camp and the shrieking emotionalism of the extreme antivivisectionists. In any case, I was looking forward to meeting Rollin. I'd never seen a photograph of him but was sure I knew what he'd be like. The typical animal-lover sort. Vegetarian, cruelty-free (no leather, no cologne), gentle, soft-spoken, with a whiff of Gandhiesque enlightenment.

Rollin responded to my knock on his office door so quickly, and with such vigor, that I thought the door might come off in his hand. His enormous, hairy hand. "You're the raptor lady!" he called from behind a mass of frizzed facial fur, knowing that I was fresh from interning as a raptor rehabilitator in Vermont. He grabbed my hand and crushed it, then yanked me through his office door. "Aren't those East Coasters a bunch of fuckin' tight-asses?" Rollin sat in the only chair and motioned for me to sit down, too.

"Um," I said brightly, searching for a place to sit, or even stand comfortably, in the disarray of papers, files, and books, which I thought might conceal some esoteric organization known only to its owner but which I discerned over time had no organization whatsoever. I spotted a stack of books that looked about chair height, perched lightly upon its edge, and began to take in the details. Professor Rollin is big. Not tall, but big. He lifts weights and eats steak. The day I met him he was wearing a tight black T-shirt that pictured a smiling skeleton riding a Harley. He had silver skull rings on his fingers, and a huge belt buckle with more skulls. I learned quickly that he possesses a sharp, refreshing, arrogant intellect.

Regarding the human responsibility to take animal consciousness into account when acting in either the domestic or the wild sphere, my views at that time were, and in fact remain, quite lofty, nonutilitarian, and hardcore. But concerning purity of intent, and of action, Professor Rollin taught me a

tremendous lesson. He has been sharply criticized by aboli-
tionist animal rights organizations — those that feel *any* use
of animals in science or agriculture is inappropriate. By working to improve the conditions under which such animals live,
these organizations believe, Rollin is giving tacit approval to
the continued use of animals in these spheres.

Rollin's response? "I refuse to go down in a flame of glory,
screaming about the best reasons for doing things and accom-
plishing absolutely nothing." He made this proclamation over
and over in the years that I worked with him, waving his arms,
punctuating with expletives, and pointing at the sky. "We are
a long way, as a culture, from abolishing the use of animals in
farming or experimentation. We're just not there," Rollin
states, rapping us straggling dreamers on the head, "and it's
not going to happen in the near future, so get your brain out
of the clouds. Human research and agriculture involves hundreds of thousands of animals, and we can help them, or we
can let them rot while we hold fast to our philosophical purity." Rollin has improved conditions for veterinary school animals across the nation, working to require adequate pain
medication, and limiting the number of surgeries that might
be practiced on a particular animal. He has worked for larger
enclosures for farm animals, and even for pain medication for
animals undergoing castration. (Rollin's joke: A farmer was
castrating his horses by crushing their testicles between two
large stones. "Does it hurt?" asked a bystander. "Not if you
move your thumbs," replied the farmer.) The abolitionists
haven't abolished anything.

Curiously, the major setback for Rollin's do-less-harm
program has been getting scientists to engage in a discussion
of animal ethics at all. The parameters of mainstream science

are not easily overstepped. The common (but philosophically untenable) claim that science is value-free is deeply entrenched, and there is a reliance on empirical observation as the source of truth — or, more properly, *fact* — that would seem to disallow the consideration of anything as murky and subjective as another creature's "mind," let alone the ethical responsibility such an attribution would imply.

Yet Darwin, the revered father of modern biological understanding, was wonderfully at home in these spaces — between the things that we observe and those that we surmise, between what we can know and what we can intuit, between what we can control and what we can simply watch. And it is in these spaces, so often, that we surpass simple fact, entering a place of wilder, more difficult, more complex truths.

While Darwin's sensibility regarding a continuity between the consciousness of human and nonhuman animals is surely in line with an emerging evolutionary consciousness, it is telling that this continuity formed the cornerstone of his naturalist's sensibility early on, before he had any awareness of his lurking evolutionary tendencies. It is one of the key traits that allowed his rapid ascent in the world of biological ideas. Darwin felt, and enjoyed, his own residence in the creature world. From almost the very beginning, he refused the arrogance of human separateness, allowing himself an entrance into wild life, a unique communion with the natural world that gave his intelligence broad play. In Darwin's rare watching, every creature is allowed to stand as each of us stands — both as a distinct individual, and with our edges blurred, in flowing lineage.

It is here that the beautifully counterrevolutionary nature of Darwin's thinking finds its nexus — in terms of science

and philosophy then and now. In the nineteenth century anthropomorphism was not uncommon in natural history reporting. But it was just that: the projection of a human glow
upon the behaviors of animals, an assumption that when animals seemed to be happy, or to be suffering, or to be curious,
they were approximating a higher human version of the same
sort of feeling. Humans retained a privileged status, for both
scientists and clerics, as the crown of Creation.

But according to that definition, Darwin wasn't really an-
thropomorphizing. In his observations of seals and caracaras
and ovenbirds and earthworms, in his records of their behaviors and his sensing of their thoughts, he utterly, and
even joyfully, abandoned his privileged human status. He
threw his own thoughts and behaviors right into the animal
mix, putting all creatures, including humans, on the same continuum of consciousness. Rather than imposing human concepts upon animal behaviors, he animalized consciousness in
general. The human "privileges" imparted by advances such as
language grew out of this continuum rather than being
plopped down on top of it. With this mind-set, very sure in
Darwin, he embraced a new kind of humility — a radical humility that made him strangely slender, able to peer into the
wilderness through doors only slightly ajar, to see, in a new
and small and gracious way, the movement of life.

Over time, this understanding would result in an evolution
of Darwin's own behavior. It was quite early in the voyage that
Darwin lost his taste for the bloody, purely sport-driven hunting of birds that characterized his youth. He had, after all,
plenty of opportunities to kill wild creatures for the sake of
his scientific collections. In fact, this was necessary in the world
of nineteenth-century science, and Darwin collected with no
more abandon than most naturalists loosed in the tropics. But

as the voyage wore on, he tried to do
less of the collecting himself, handing
over the task to local boys or Syms
Covington. It was not only a matter
of claiming time for his field work
and record keeping but also a change
in taste. The more he came to know
wild animals directly, the less he felt
able to collect them with his earlier
enthusiasm. This might seem obvious

ANDEAN CONDOR

to us; if we love animals, we don't go about shooting them out
of the treetops just to get a look at the contents of their intestines. But for a nineteenth-century naturalist, such a sentiment
was singular. He was still obliged to collect animals himself,
but as time wore on, he did it as little as possible.

Rather than toughening Darwin up, his deepening scientific knowledge left him more and more sympathetic toward
creatures. The examples in his notes are many, and most of
them are not particularly remarkable but reflect a simple, normal, humane sensibility. He was taken aback and sickened, for
example, to hear of cruel experiments conducted by Chileno
countrymen to show "that the condor will live & retain its
powers between five & six weeks" without eating. Most of us,
too, I am sure, would be willing to call up similar compassion
for a large bird — even a carrion eater such as the Andean
Condor, toward whom the human psychological reaction
tends to be complex. But it is less common for human sympa-
thy to extend to the further reaches of the phylogenetic scale,
as Darwin's did.

In the *Journal of Researches*, Darwin tells the story of a wasp
and a spider observed in a Shropshire hothouse. A large female wasp was caught in the web of a small spider:

This spider, instead of cutting the web, most perseveringly continued to entangle the body, and especially the wings, of its prey. The wasp at first aimed in vain repeated thrusts with its sting at its little antagonist. Pitying the wasp, after allowing it to struggle for more than an hour, I killed it and put it back in the web.

Darwin was, and remained through the whole of his life, the sort of person who would actually intervene in natural processes to keep an insect from extended suffering.

In his autobiography, a strange, slim volume composed late in his life, Darwin chalked it up to his becoming more civilized in general. "I discovered," he wrote, "though unconsciously and insensibly, that the pleasure of observing and reasoning was a much higher one than that of skill and sport. The primeval instincts of the barbarian slowly yielded to the acquired tastes of a civilized man." Like the rest of his autobiography, these lines make Darwin seem much duller than he actually was. He does not mention that his move from a barbarian to a man of civility involved lying on his belly to look ovenbirds in the eye, swimming with iguanas, and singing to earthworms. He seems to have forgotten that the "acquired tastes of a civilized man," at least in his case, encompassed the delicate ability to raise firefly larvae in tiny glass dishes by hand.

Darwin was captivated by the fireflies of Rio, insects in the family Lampyridae, the same family as the English glowworms. The going theory regarding their stunning luminosity had to do with sexual attraction. The insects would glow so that the males and females could find one another in the darkness. But in Rio Darwin observed the larvae closely and no-for humility, but I believe arrogance is, rather, a symptom of humility's true opposite — forgetfulness. And it is in noticing these moments of surprising forgetfulness that we renew humility, a remembrance of the ground of our lives, the basis of what is offered to us as well as what is required of us.

So often we are not ready for such sight, because we know too much, and are too impressed by what we know. As scientists, as teachers, as naturalists, as parents we scatter in wild places, brandishing our field guides, the radius of our vision set, ready to impose our astonishing knowledge. There are myriad ways we proclaim our love of wild things, even as we trample them underfoot. I know about this — I do it all the time. Refuge in our own knowledge is perhaps our favored species of forgetfulness.

For Darwin, the rhea story became more significant in the telling and retelling; even so, I realize I am imbuing it with a layer of meaning that Darwin might not have intended. In my story, Darwin becomes more careful after his experience. It is very likely true. I wonder, in measuring himself against d'Orbigny, whether Darwin's own judgment was clouded, whether in petty competition his best sense of himself as a naturalist went missing. Was it in such a cloud that the Lesser Rhea was nearly lost? In my telling of the story, Darwin's near-eating of the stewed rhea was humorous, but it also snapped him back to more lucid thinking. The mishap helped to root him out of worries over d'Orbigny and back to the truth of his subject. His future observations would be subtler, filled with even more of the object and less of himself. They would be more mindful and, over time, humbler.

It's a lovely word, *humility*, rooted in the Latin word *humus*, meaning "soil," "land," "the earth itself." In learning and re-learning humility, we are reminded of our creatureliness, our ticed that they glow, too, though more feebly than their parents. Larvae don't reproduce with one another, so either their luminosity was superfluous, or it existed for some other reason. Darwin did not solve the mystery of lampyris luminousity, but in order to study the larvae carefully, he "kept some of them alive for some time," noticing that when touched even slightly, they "feigned death and ceased to shine."

And how did this young man with large hands care for his miniature larvae? "I repeatedly fed them on raw meat," he recorded in the *Journal of Researches*, unaware of the charm in this confession. For surely this is what we all want in a companion — someone with the patience and know-how to raise something as tiny, ugly, and helpless as a lampyris larvae. I picture Darwin peering over their little dish, proffering the gift of meat — patience, gentleness, and relentless curiosity all mingling happily. (Before we begin the preparations for Darwin's beatification, however, I feel obliged to point out that he discovered that the adult insects remain bright beyond death by yanking their little heads off and watching them glow.)

In spite of the practicalities that faced a working biologist in the nineteenth century, Darwin's easy participation in animal lives did not subside in a life that became, after his return to London, increasingly distant from things wild. When he wasn't busy putting wasps out of their misery, he was working with his wife, Emma Wedgwood Darwin, to ban steel leg traps in the capture of "vermin." He was known to jump out of his own carriage in order to upbraid someone who was whipping or excessively spurring a horse, and when he became a local magistrate, he delivered the stiffest possible fines for the mistreatment of farm animals. Eventually, he was drawn into the highly complex debate over vivisection and walked the line delicately, if not as bravely as he might have, opposing cruel or useless animal experimentation while hoping to reserve the rights of scientists to work in their own homes without government intrusion. In the title of his popular book *The Expression of the Emotions in Man and Animals,* published in 1872, he, as Bernie Rollin puts it, "brazenly hoists a middle finger to the Cartesian tradition, since Darwin saw emotion as inextricably bound up with subjective feelings."

The extent to which an appreciation of such subjectivity pervaded Darwin's entire life and thinking is revealed in the very last of his works, a book with a title that appears designed to deter the faint of heart, *The Formation of Vegetable Mould Through the Action of Worms with Observations of Their Actions,* published in 1881, less than a year before his death. Earthworms. The subject was entirely unexpected by Darwin's colleagues and his wary publisher, but the book turned out to be wildly popular and today remains, after the *Origin of Species* and *The Descent of Man,* Darwin's bestselling title. Here, he spent more than three hundred pages marveling in expansive detail over the work of earthworms as they denude the land, enrich the soil, and carry out all manner of geological and archaeological projects. In his conclusion Darwin cheerfully informs the reader, "In many parts of England a weight of more than ten tons of dry earth annually passes through their bodies and is brought to the surface in each acre of land; so that the whole superficial bed of vegetable mould passes through their bodies in the course of every few years." I remember dissecting earthworms in seventh grade and finding little more than a tube within a tube, the simplicity of their structure belying — or, rather, allowing, I suppose — the complexity of their ecosystemic duties. The worm book sums up Darwin's life work so well: in the accumulated action of the very small lies the axis of earthly existence.

Although consideration of earthworm consciousness was beyond the professed scope of his project, Darwin could not help himself. "As I was led to keep in my study during many months worms in pots filled with earth," he confessed, "I became interested in them, and wished to learn how far they acted consciously, and how much mental power they displayed." Besides, no one else was bothering to pay the minds of worms any mind at all. It was clear enough that worms are "poorly provided with sense-organs," they "cannot be said to see, they can just distinguish between light and darkness; they are completely deaf, and have only a feeble power of smell." Darwin was eccentric, but not unreasonable. Even he could see that these blind, wiggly little tubes weren't doing much thinking. So Darwin was surprised to find that "they should apparently exhibit some degrees of intelligence instead of a mere blind impulse." This stemmed, he thought, from their one highly developed sense — that of touch. Using their skin to gather information about their environment, Darwin's earthworm officemates demonstrated some skill in the creation and lining of their burrows with leaves and castings and also in plugging their entrances. They did, in fact, "act in nearly the same manner as would a man," sifting through various kinds and shapes of materials, transporting them in different ways depending on what they had chosen.

Darwin's years of earthworm observation led him to believe, further, that worms enjoy a measure of simple pleasure (rather than "pleasure"), particularly in the consumption of favorite foods, in sexual activity, and in very basic social relations — preferring to lie next to one another at times, for example, rather than alone. When engaged in such activities, the worms did not respond reflexively, as they usually did, to obnoxious stimulus such as a bright light. When higher animals disregard a stimulus to which they would normally respond, Darwin reasoned, "we attribute this to their attention being then absorbed; and attention implies the presence of a mind." Darwin knew he sounded crazy. "The comparison here implied between the actions of one of the higher animals and of one so low in the scale as an earth-worm may appear far-fetched, for we thus attribute to the worm attention and some mental power," he admitted, even as he forged ahead. "Nevertheless I can see no reason to doubt the justice of the comparison."

According to the yardstick of human intelligence, worms do not rate. They cannot see, or hear, or vocalize. But Darwin's studies show us the limitations of what we take for granted as proper measure. We find ourselves repeatedly astonished by animals' own revelations about themselves. Parrots learn with their tongues, worms learn with their skin, pigeons navigate by the earth's magnetism, by the moon, by the stars. Who would have thought? Even as we come to understand more and more, animals will continue to possess their own minds and ways that lie just outside our imagining. The naturalist's faith draws us into the work of recognizing such dimensions of nonhuman lives — both what we can know and the pure subjectivity that lies, however frustratingly, beyond the reach of our brilliant science. We are invited to stand as witnesses, rather than all-seers, to give all creatures space for the breadth of their secret knowing.

"Come, little friend," Darwin spoke to the tiny, ground-stepping ovenbirds of Peru, traversing the human-animal gap so lightly. And "little friend," here, is not merely an affectionate term; it also invokes Darwin's own understanding of the potential to enter into relationship with nonhuman creatures. We can, for starters, bring one another into our spheres of vision, respond to one another in various ways, and, from the human side at least, accomplish this appropriately or not. We can come to recognize, as Darwin did, that the stories told by individual animals may have a place in biological science — that without these voices, our science becomes diminished and somehow less *true*, even as it advances. *Bien te veo,* we can hope, someday, to say truthfully. *I see you well.*

CHAPTER SEVEN

Condors Flying

The Condor is well known to have a wide geographical range, being found on the west coast of South America from the St. of Magellan throughout the entire range of the Cordillera. On the Pagagonia shore, the steep cliff near the mouth of the Rio Negro in Lat 41 was the most northern point where I ever saw these birds or heard of their existence. They have here wandered about four hundred miles from the great central line of their habitation in the Andes.

— *Ornithological Notes,* 1834

*D*arwin found the cold grasslands of southern Patagonia to be singularly monotonous. "The country remains the same, & terribly uninteresting," he complained in his diary as the *Beagle* began its eighteen-month exploration of South America's southwest coast in the spring of 1834. It was becoming clear that the journey would be much longer than expected, and Darwin felt both his spirits and his scientific resolve beginning to flag. It was not just the landscape but also the flora and fauna that bored him, impatient now for a renewal of the inspiration he felt in Brazil and finding it nowhere. "The great similarity in productions is a very striking feature in all Patagonia, the level plains of arid shingle support the same stunted & dwarf plants; in the valleys the same thorn-bearing bushes grow, & everywhere we see the same birds & insects." There were "ostriches" now and then, but they were wild and unapproachable. Darwin wasn't really surprised. The undersized plants and unfriendly cactus didn't seem fit for much life at all. He had always refused to apply himself to botanical study, feeling that there were just too many plants to master, and that others could take care of such things better than he could. So as for the sharp yellow tussock grasses, the strange scattered ferns, the enormous terrestrial bromeliads and their starry flowers, the cacti like tiny trees, none of these seemed to interest him or even strike him as noteworthy as he trekked this highly unique ecosystem. He was lagging a little by this time, tired, homesick, his immune system depressed, and weeks away from a bout of typhoid that would rend his body while mysteriously reviving his spirits.

There was one animal, though, that was large enough, plentiful enough, and odd enough to drag him a bit from his doldrums and inspire him back into making some decent naturalist's notes. The guanaco, a large desert mammal, was found in great herds in southern Patagonia. It is a distant relative of the camel, a near ancestor of the llama, and between the two in size. It is long-necked like a llama, with the same sort of camel-ish face and the same look in the eye, as if it might spit at you if you glanced at it sidelong or wore some fashion accessory of which it did not approve. Guanacos are oddly proportioned, with bodies that seem too wide and too heavy for their slender legs and ballerina toes, and necks that seem too long for any practical purpose at all. Certainly they don't have to reach for leaves as a giraffe does, since, as Darwin so bleakly noted, the landscape bore no trees.

The physiological adaptations that would allow such a sizable mammal to sustain itself in the tropical alpine deserts are little understood, particularly the guanaco's ability to conserve water. Darwin had read that they were sometimes seen drinking from brackish pools, and his own observations confirmed this unusual fact. Though more study is needed, biologists now suspect that guanacos may fall into a kind of torpor during the freezing nights, with their body temperatures falling to keep from using excessive metabolic energy, then rising dramatically during the heat of the day, as the body temperatures of camels do, to avoid loss of water through perspiration.

Darwin was tickled by the guanacos. They broke up the landscape in large herds of fifty or one hundred, and in one instance he probably exaggerated only slightly in suggesting that he'd seen a group of five hundred. They behaved strangely, a little eccentrically, even. He had seen a guanaco, "when disturbed, not only squeak or neigh, but jump & prance in the most ridiculous manner, apparently in defiance as a sort of challenge." This was just Darwin's sort of beast. He found great heaps of guanaco dung, eight feet in diameter, and learned that the animals sleep in a circle with their heads facing outward, "to keep watch for the Lions, & hence the dung." Not only did the guanacos prefer to defecate where others did, but also to die where others had. Darwin noticed "bushy places near the river" that were "white with bones." Wounded guanacos would go to the river and lie among the bones of their guanaco colleagues, dying where their skeletons could not be "torn by pumas." Darwin was impressed with the guanaco's kittenlike curiosity, "for if a person lies on the ground, & plays strange antics, such as throwing his feet up in the air, they will almost always approach by degrees to reconnoiter him." Darwin fails to mention just how he came up with this little tidbit — was it his own waving toes that the guanacos "reconnoitered?"

At Rio Santa Cruz, Darwin's party discovered a dead guanaco in shallow water. The meat appeared fresh, and upon skinning the head they discovered a bruise. "We imagine that the Indians must have struck it with their balls," Darwin reported in his diary as if contemplating the scene of a crime, and "that going to the water to drink, it died." Whatever the guanaco's end may have actually been, the corpse was "after a few doubtful looks, voted by the greater number better than salt meat, & was soon cut up & in the evening eat." Darwin does not mention whether he was, himself, "in the greater number," but two days later he alluded again to a general hesitation before the idea of consuming meat that was discovered dead rather than expressly killed for eating. "A Guanaco was shot," he wrote, "which much rejoiced those who could not compel their stomachs to relish Carrion."

And who were these squeamish sailors who could not quite manage to cheerfully consume perfectly good guanaco? The deeply ambiguous relationship between humans and carrion is curious, even within this late Georgian–early Victorian culture that lived much closer to the origin of its food, a culture in which the connection between death and sustenance and continuation was much nearer the skin. Jane Austen herself, like her middle-class characters and like Emma Darwin, would have walked between slabs of meat in the cold cellar — slabs still shaped forthrightly like the lambs and cows and birds that so recently enjoyed the near pasture — helping the cook decide on the proper cut for the evening meal. The dead bodies of loved ones — mothers, children, husbands — would linger in the home, not having been rushed away for a proper modern death in a sanitized hospital bed, and would be prepared for burial by the women in the family, then put on view in the parlor for all to see. In these same rooms where the bodies were washed, births would eventually follow.

Yet carrion, an animal lying there after a natural death, was, in spite of its perfect practicality as an ingredient for stew, somehow unsettling. It is as if a thing not explicitly killed for the purpose of eating was somehow *more* dead, or perhaps more brazenly bearing another dimension of death's truth — its utter separation from human control. There is an awkward stillpoint here in these animal bodies that are flesh yet not quite meat.

As the members of the *Beagle* expedition raised their hands in favor of dining on dead guanaco, or did not, the Andean Condors began to drop from the sky. They had been soaring so high there that no one had even seen them. Condors are among the only birds on earth with enough mandibular strength and cervical muscle to rip through the fresh hide of a mammal as large as a guanaco. This they accomplish with apparent gusto, bracing with their thick legs and pulling strongly with their necks and shoulders, before plunging their bare heads straight into the animal's intestines. Once he began to watch the condors, Darwin couldn't take his eyes off of them. During his time on the west coast of South America, in his travels from Patagonia to Peru, he lavished time and attention and thought upon these birds. Excepting the rheas, Darwin wrote more in the *Ornithological Notes* about Andean Condors than about any other single bird.

Darwin observed that the condor's presence in a place "appears to be chiefly determined by the occurrence of perpendicular cliffs." In Patagonia they would both hunt and breed from these same cliffs, while in Chile they "haunt the lower Alpine country nearer to the shores of the Pacifick, & at night several roost in one tree." But, he added with seeming appreciation, these Chilean condors will retire in early summer, "to the most inaccessible parts of the inner Cordillera, there to breed in peace." As usual, Darwin peppered his own observations with intelligence gathered from native observers. He was assured that young condors "cannot fly for the first whole year." And he himself had seen a young bird at Concepción grown nearly the size of the adult beside it, but "covered over its whole body by down precisely like that of a Gosling." (Well, he found it necessary to clarify, the down was "of a blackish color," but "*otherwise* just like that of a gosling.") Darwin felt sure that this bird was already many months old and would not be able to use its wings for many months more. Actually, most Andean Condors fledge when they are about six months old. Darwin probably underestimated both the rapidity of a young condor's growth, and the speed with which down can be replaced by workable feathers once the process is under way.

Still, the condor's protracted breeding cycle is unusual among birds. A young condor won't breed until it is six years old, and its white feathered neck ruff appears at this time. Watching a pair of condors fly with a young bird, one that did not yet have its ruff, Darwin remembered the gosling bird that he had seen just one month earlier. He rightly conjectured that this flying young bird must have been born the previous summer, at least, and that condors probably breed only once every two years. This slow reproduction rate is one of the primary reasons that the Andean Condor is so susceptible to human encroachment and persecution. Its numbers are declining steadily. "It was a fine sight to see between twenty or thirty of these great birds start heavily from their resting place, then wheel away in majestic circles," Darwin wrote dreamily, lines for modern naturalists to sigh over. Andean Condors are rarely, if ever, observed in such numbers today, and it is unlikely that they ever will be again.

Darwin learned from the "Chileno countrymen" that condors frequently attack very young goats and lambs. Shepherd dogs were trained to respond to a condor overhead by running about while looking up and barking violently, in hopes of chasing it away. The Chilenos themselves caught and killed as many as they could, which was not as difficult as one might guess given the imposing nature of the bird. Two methods were divulged to Darwin. In the first one, the carcass of some sort of livestock was placed within an enclosure in an open area, and after a group of condors had discovered it and gorged themselves, men would gallop in on horseback and gather them up, "for this bird, not having space to run, cannot give its body momentum sufficient to rise from the ground." This is true, and after a particularly large meal, condors often have a difficult time getting off the ground at all, whether they are surrounded by horses or not. The second method of condor capture was even more amazing. The condor snatchers would "mark the trees in which [the birds] roost, frequently to the number of five or six together, & then at night climb up & noose them; they are such heavy sleepers, as I have myself witnessed, that this is noways a difficult task." Those that weren't killed were kept or sold as ill-fated "pets." In an unshakable image, Darwin wrote that "several hours before any of the condors die, all the lice with which they are infested crawl to the outside of the feathers." He collected a specimen of the lice, and labeled it *Ricinus,* a common avian ectoparasite.

As Darwin became a more seasoned watcher of condors, the behavior that intrigued him most was their mystifying ability to congregate, gathering themselves together from unseen distances, about the body of a guanaco. "When an animal is killed in the country," he explained in the *Notes,* "it is well known that the condors, like other carrion vultures, gain the intelligence & congregate in an inexplicable manner." Great numbers of condors would gather around the fresh carrion — a guanaco that had died of "natural causes" or been killed by a puma. A sky that seemed still and empty would suddenly be filled with condors, all spiraling toward this single point, and Darwin was baffled by it. Surely one or two birds flying overhead might have seen the animal, but where did the others come from? Condors are nearly silent, not communicating with others vocally, so they couldn't have "spoken" to one another about the find. Could they have sniffed it out? How well could they smell?

The power of the olfactory sense in the vulture family has long been, and in fact remains, controversial. In Patagonia, Darwin recalled the work of John James Audubon, whose experiments on the olfactory capabilities of vultures and "such birds" were published in 1826 and concluded that these birds, for all practical purposes, had no sense of smell. In his series of feeding trials, Audubon offered Turkey Vultures foul, putrefying, ill-smelling meat, which the birds ignored. Surely, if they could smell, this good rotting stuff would attract them, Audubon surmised.

But thinking that condors might have capacities that differ from their smaller relatives, and still wondering over their mysterious congregations, Darwin decided to take matters into his own hands. In a garden where captive condors were kept, each tied to a wall by a rope in a long row, he concocted this good experiment:

> Having folded a piece of meat in white paper I walked backwards and forwards, carrying it in my hands, at the distance of three yards. No notice was taken, I then threw it on the ground within one yard of an old cock bird, he looked at it but took no further notice. — With a stick I pushed it closer; the condor at last touched it with his beak, & then instantly with fury tore off the paper, at the same moment, every bird in the long row was struggling & flapping its wings.

In the same situation, Darwin was sure, it would have been "impossible to have deceived a dog" — the dog being one of his favorite compasses in making important comparisons (when he was considering marriage, he wrote up a now-famous pros-and-cons list, which concluded in the pro column, "Better than a dog, anyhow"). Darwin rarely conducted actual experiments on birds, and this one is a testament to his fascination with condor lives. Think of the young Charles Darwin walking in a garden full of condors, birds that stand as tall as his waist, waving his meaty parcel. Such an image. Clearly condors could barely smell at all.

Yet when he returned to London Darwin was further confused by Cambridge professor Richard Owen's presentation before the Zoological Society concluding that Turkey Vultures "must possess an acute power of smelling," an assertion Owen supported through an astute and detailed examination of the birds' olfactory nerves. Darwin reported all of this — Audubon, Owen, the meat-ripping garden condors — in his

1838 *Journal of Researches* and concluded, as confused as ever, that "the evidence in favour of and against the acute smelling powers of carrion-vultures is singularly balanced."

To make matters worse, the "smelling powers" of vultures have somehow fallen into the realm of modern ornithological myth — the sort that is actually perpetuated by ornithology texts and teachers. (Another such myth is that cormorants are so primitive that they lack oil glands, are thus unable to waterproof their feathers, and so must spread their wings to dry. Cormorants do spread their wings to dry, but they have perfectly functional oil glands.) When I began interning as a young raptor rehabilitator, the director of the center where I worked duly and solemnly reported to me that our captive Turkey Vultures, like all vultures, could not smell. This, she went on, was an adaptation to allow them to eat putrefying flesh without being put off by the odor. I believed her and repeated the myth to unsuspecting schoolchildren.

But modern ornithologists are finally disentangling the threads of the vulture-family olfactory mystery, at least somewhat. For starters, it turns out that the entrenched belief that vultures are somehow attracted to putrefying animal flesh is mainly false. It is true that birds in the vulture family are able to resist botulism and withstand bacterial toxins well beyond the norm. The meat they eat is often far from fresh. But like most animals, vultures avoid decomposing flesh if they possibly can and sensibly refuse meat that is in the late stages of putrefication. The belief regarding vultures' being happily unable to smell their rotten food is just as false as the conclusion of Audubon's putrefying flesh study. They ignored his disgusting meat not because they couldn't smell it but because they didn't want it.

There is one more strange little turn in this story. For

decades nearly everyone assumed that the question of vulture-family smelling could be solved by studying a single species and then extrapolating to the others. In the case of closely related species, that would normally be a reasonable assumption from which to proceed. But in this case, much of the confusion centers on the quite recent realization that species within the family are, in this respect, entirely different. It turns out that while Richard Owen has been vindicated by modern studies — Turkey Vultures can actually smell quite well, for a bird — Darwin was also right. Condors cannot smell.

This phenomenon is difficult for a modern ornithologist, steeped in evolutionary theory, to understand. Surely it makes sense for a carrion eater, a bird living a sometimes marginal life entirely dependent on proficiently finding dead meat, to be able to smell. The going theory, sensible but unproven, is that the increased brain capacity required to support a strong olfactory sense would result in a heavier brain (and hence skull and head) that would be too costly for the condor's life of high, sustained flight, and that this disadvantage outweighs the potential benefit of smelling food.

But the young Darwin was not armed with insights such as these, insights that his own eventual theories would ground. Alone with his condors in Patagonia, and not knowing what else to do in order to unravel the mysteries of condor existence, he stood staring at the sky, where condors turned and turned in effortless circles without, it seemed from the earth, ever lifting a feather, and at an altitude at which Darwin had little imagined a being of flesh and bone could possibly survive. Darwin was transfixed by this flight, and he would lie on his back and watch condors in the sky for longer than it seems reasonable to watch almost anything. He lost himself in the watching and let these high, wild beings lift him out of the dreary landscape and into the wisps of the clouds. It is no wonder Nora Barlow called the Andean Condor "Darwin's great bird of the imagination."

He watched flying condors in Lima for nearly an hour, he claimed, without once taking his eyes off of them. "It is beautiful," he wrote simply, disarmed, "to observe their manner of flight." Darwin was amazed at the condors' ability to sweep in large, curving circles, to ascend and descend and to do this "without once flapping" for as long as he could bend his neck back to watch them. Several birds glided close over his head, and he watched, "from an oblique position, the separate & terminal feathers of the wing; if there had been the least vibratory motion, the outlines would have been blended together, but they were seen quite distinct against the blue sky." Even in flight, the condors were strangely motionless.

Eventually, Darwin would try to explain what he saw, try to account in some way for the marvel of condor flight:

> In the case of any bird soaring, it must have sufficient rapidity of motion, so that the action of the inclined surface of the body on the atmosphere may counterbalance its gravity. The force to keep up the momentum of a body moving in a horizontal plane in that fluid (in which there is so little friction) cannot be great, & this force, is all that is wanted. The slight movements of the neck and body of the condor we must suppose sufficient for this.

Well, it was a gallant effort, given that Darwin lacked the benefit of modern flight physics (though that bit about the head and the neck somehow steering the condor and keeping it aloft does go rather far astray). We now better understand the role of thermals, rising currents of warm air (and also, in the condor's mountainous habitat, the currents that sweep up the windward sides of hills and cliffs), in lifting the birds and allowing sustained flight that is, truly, almost effortless. The head motions that Darwin described are more likely the simple movements of a bird seeking food or just surveying the view as it soars and probably have little to do with momentum. Steering is accomplished with almost imperceptible movements of the flight feathers on the outer wings and tail, and very subtly by the alula, a small, strong, dexterous feather on the bird's wrist, where the major wing bones meet.

Darwin would have delighted in knowing that these New World vultures descended from birds that were true predators — catchers of live prey — and became exclusive carrion eaters over the slow course of evolutionary time. This gliding flight that he admired was the trading point. It is surely much easier to be a predator, to be able, if you are a hungry bird, to go out and kill something rather than waiting for something tasty to die. But of course if there happens to be a good bit of dead meat on the ground, it requires less energy to flap down and eat it than it does to hunt and kill something. In the transition to carrion eating, vultures could eat larger meals in one sitting. They became bigger and heavier, gaining wider wings to sustain them in a soaring flight better suited to seeking carrion but losing the speed and maneuverability required by a predator. Adaptations were refined, right down to the tiny webbing between a condor's feet used as resistance in steering and landing, until the vultures in general, and the condors in particular, became the most energy-efficient beings that soar over land. In their adaptations for sustained soaring they are rivaled only by the pelagic albatrosses and frigatebirds. It is wonderful to see the continuum from Darwin's early musings on these birds before he himself understood, or even accepted, evolution at all to our deeply evolutionary understanding of such creatures, founded very soundly on natural selection.

The theory Darwin eventually formulated regarding the condor feeding congregations was probably close to the truth:

> Where the country is level I do not believe a space of the heavens of more than 15° above the horizon is commonly viewed with any attention, by a person either walking or on horseback. If such is the case & the Vulture is on the wing at a height of between three and four thousand feet, before it could come within the above range of vision, its distance in a straight line from the beholder's eye would be rather more than two British miles. — Might it not thus readily be overlooked? When an animal is killed by a sportsman in a lonely valley, may he not all the while be watched from above by the sharp-sighted bird? And will not the manner of its descent, proclaim throughout the district to the whole family of carrion feeders, that their prey is at hand?

The native people supported this theory, yet Darwin would always remain insistent that while condor flight might often have this practical end, there are times that condors fly just to be flying, just because they do it so well and so beautifully. He watched them turn "in the most graceful spires & circles," he wrote in the *Notes*. "I feel sure they do this for their sport" or, as he put it in the *Journal*, their "pleasure."

Though Darwin worked to understand the flight dynamics of these birds, it was clearly not just the astonishing fact of condor flight — its altitude, its stillness, its grace — that gave Darwin such pause. It was the coupling of such flight with the ground from which it rises: bare, dull, brown, dry ground, whose monotony is broken only by living guanaco, and then dead guanaco upon which these same condors inelegantly feast. Other scavenger birds attend guanaco kills as well — both the smaller vultures and the caracaras. But it is the condors that begin the feeding, opening the animal for the smaller birds. And it is the condors who stay until the task of cleaning the skeleton is complete, for in spite of their size and the strength and thickness of their mandibles, the condors possess bills that resolve into tiny, tweezing forceps, delicate enough to glean bits of muscle and tendon from between the tight skeletal articulations. What begins in a rather bloody, tearing fury, attracting a hoard of scavenging birds, ends with the condors alone, calm and meticulously nibbling, almost artful in their effort to leave nothing at all but a bare, new skeleton.

Darwin watched the process from start to finish. He was struck by the image of the smaller vultures and caracaras standing within the ribs of a guanaco, and he mentions it several times in his writing. It's an affecting scene: the bare landscape, the guanaco, now mostly a skeleton, the creatures that fed upon it standing there, framed by its curved bones. Darwin was standing there, too, stroking his beard, then perhaps feeling his own ribs with a new and brisk awareness of his own bones covered — more thinly than he had realized — with flesh, a flesh that is elemental, near to the earth, and near also to these condors with their featherless heads (for feathers would only become filthy in their feeding). And then in these birds this same flesh is borne up, farther than seems possible, rising without effort.

Ikkyu was a fifteenth-century Buddhist poet and monk, prominent in the history of Japanese Buddhism. The violence of the protracted Onin civil war was the backdrop for

Ikkyu's life and his writings. From his youth to his old age he watched the sacred buildings in Kyoto reduced to shards and the people around him, both rich and poor, living in ruin. Ikkyu's *Skeletons*, a string of prose and verse with inky illustrations in the middle, was translated into English by R. H. Blyth in the 1960s. The writing is instructive, but it is the drawings that are most famous. In them, little skeletons are busy about their day, playing the flute, having a temper tantrum, tending the sick, making soup, making tea, making love. The skeletons are active and good-humored, and they are, at first glance, more cute than eerie. In a couple of images, there is a man with flesh on his bones, wearing a *hakama*, traditional Japanese dress. A skeleton greets him, grinning. "The beginner must do *zazen* earnestly," Ikkyu instructs at the start of his manuscript. "Then he will realize that there is nothing born into this world which will not eventually become 'empty.'"

Yeah, yeah, we're all really skeletons. Didn't we all learn this in the existentialist segment of our sophomore introduction to philosophy? But Ikkyu's angle is different, both didactic and somehow celebratory. Ikkyu falls asleep in the temple, and in his dream he sees many skeletons assembled, "each moving in its own special way, just as they did in life." One of the skeletons approaches Ikkyu and speaks to him in verse.

For as long as you breathe
A mere breath of air,
A dead body
At the side of the road
Seems something apart from you.

And when is such thinking not a dream, Ikkyu wonders upon waking, who is not in truth a skeleton? "When the breathing stops," he muses, "and the skin of the body is broken there is no more form, no higher and lower. You must realize that what we now have and touch as we stand here is the skin covering our skeleton. Think deeply about this fact." No higher and lower, no separation of the earthen and the transcendent. Ikkyu tells us that he and the skeletons went on to enjoy one another's company quite a lot.

Ikkyu's skeletons invite us into that ground of being where, I think, Darwin found an unusual comfort. He had never quite known before that earthen grace where, like the plain space between a guanaco's ribs, a bird can stand. "We go back to our origin, we become earth again," Ikkyu wrote, knowing it was nothing new, nothing to be shocked by even then.

By fall of 1834, Darwin and several shipmates became very ill with symptoms that are consistent with typhoid. He had often been unwell on the voyage, but he had never been so sick for so long. What a desolate place it was to be ill, what little consolation against the awareness of one's mortality such an experience can bring. Still, Darwin pressed through Valparaiso and Santiago on horseback. "I consider myself very lucky in having reached this place," he confided simply to Caroline. "Without having tried it, I should have thought it not possible; a man has a great deal more strength in him, when he is unwell, than he is aware of . . ." Some of Darwin's most thoughtful observations on condors were put down at this time. He could watch them soar as he lay on his back, resting. Somehow, in this barren place, the renewal Darwin hoped for had settled upon him.

"Just let your body be blown along by the wind of the

floating clouds," Ikkyu wrote, hundreds of years ago. "Rely on this." Nothing in Darwin's formal studies had prepared him for a calm that could suspend his reason. He had studied condors, experimented upon them, measured them, interviewed natives about them, deliberated over their diets, their habits, their flight. But in the end he saw empty space between ribs and flesh carried to the sky. The parameters of life and thought, once seemingly circumscribed, were loosened, and the naturalist's imagination lifted. He leaned his head backward and refused to close his eyes. "Beautiful," was his summary line on condors, "and wonderful."

CHAPTER EIGHT

An Infinity of Owls

If, by means of the traps, I had not been aware how wonderfully numerous the smaller rodentia are in these open
countries, it would have been an enigma to explain the
support of such an infinity of owls.

— *Ornithological Notes*, MALDONADO, 1833

For millions of years, a small portion of the Pacific
Ocean floor called the Juan de Fuca plate has been
pressing itself, with slow, kinetic insistence, beneath
the Washington, Oregon, and California coastlines. This
movement, continuing today, is the foundation that gives
these western ocean edges their sense of unreserved wildness.
Geologically, this place is young, and its restlessness is
palpable. The coastal ranges are rising, jagged, solid, and cold.
The forests are unpredictable, with glaciers flinging out rivers
and carving high green lakes. In Washington, there lies along
the coastline a chaotic mess of drifted logs. The ancient Sitka

spruce forests grow right up to the ocean's edge, where they fall in, and then return someday, horizontal, barkless, and smooth.

The coastline is a rocky one, so the creatures at the water's edge are stone creatures; barnacles, mussels, limpets, and sea stars blanket every surface within reach of the sea's mist. Most of the many shorebirds that may be observed on the West Coast are migratory, but a couple are residents, making their homes there year-round. The largest and most conspicuous of these is the Black Oystercatcher. Its legs are fat, and its feet thick and padded for a life of walking across sharp barnacles. The oystercatcher uses its bright red bill for chiseling limpets from their rocks and for prying mussels from their navy-purple shells. The bill is specially shaped for the purpose — tall, narrow, and wedged like an oyster knife. Even so, since the shelled invertebrates have spent their entire evolutionary history contriving the means to avoid this very thing (they are glued tightly to their rocks, and their shells are strongly closed), the oystercatcher's task looks troublesome, at least. Yet the oystercatchers themselves appear unfazed and well fed.

Things look far different in the Northeast, where the coastline is known as a *trailing edge* coast, one created by geological plates that, rather than wedging together, have been drifting apart — and for a much longer time, probably two hundred million years. The East coast feels lazy to me, quiet, sandy, and soft. Walking with a dear friend at Cape Cod one autumn, I observed the Black Oystercatcher's East Coast relative, the American Oystercatcher, very easy to identify with its black head and bright white breast. American Oystercatchers walk on wet sand, picking worms and shrimp from beneath its surface. "You call those oystercatchers?" I thought to myself, irrationally possessive and proud of the oystercatchers on the West Coast, hammering their little brains out, detaching the adductor muscles of various bivalves and slurping out the shimmering centers.

One of my favorite beaches on the Washington coast, near the reservation town of La Push, is covered with basalt stones, most of them small and worn and good for sifting between fingers. Protruding among these are strange, curved basalt expanses, just wider and longer than a human body. They are wonderful to lie upon, ample and strangely soft for a rock.

There is always a mist by La Push, and even on sunny days, the ocean blows onto eyelashes and hair and the lenses of binoculars are made hazy. I sat bundled and alone on this beach one spring day in the comfortable and familiar gray damp. Just at the fog's edge I could see the scoters, guillemots, and auklets riding the waves, blithe and fresh. I lifted my binoculars to scan the large, mussel-lined rocks along the edge of the incoming tide. No birds. So I settled my back into the curve of the basalt and rested into the beautiful gray day.

Suddenly the rocks before me started shifting, breaking apart in slow motion. A dull shock shot through my bones and I senselessly grasped piles of stones with both hands and watched, wondering whether one ought to run from the ocean's edge during an earthquake — how long does it take a tsunami to brew? It took me longer than it should have to realize what was really happening, and I was nearly as surprised and disoriented as if there *had* been an earthquake. The rock had been covered with oystercatchers, settled over their pink legs with their lipstick-red bills tucked beneath their scapulars against the cold and their orange eyes closed. Now they were rising and walking, lifting their feathers, stretching one leg and one wing at the same time, as shorebirds do, and pointing their faces into the wind.

Well, this was a little embarrassing. I was supposed to be a decent watcher of birds, after all. I had written a book on the subject. I involuntarily looked around, making sure, I guess, that no one had *seen* me *not* see more than a dozen large shorebirds resting beneath my nose. But soon enough I swallowed my pride and began to laugh. One moment a rock is still, silent, solid as a stone. The next it is quivering all over, alive, enlivened by the wind, ready to take flight.

In his book *Returning to Silence,* the late Buddhist teacher Dainin Katagiri wrote something that I once found startling — startling in the fact that it had been printed for our dim western minds at all. It seemed at the time to be the sort of thing that, typically, if one had thought it up, one might happily keep to oneself. This is what he wrote: "The essence of the universe, the essence of being itself, we call the Truth, we call the universal life, common to all sentient beings — trees, birds, human beings, pebbles, rivers, mountains." This notion that the ground of existence is somehow common to all of creation is not unusual in spiritual discussions, but it is the list of what Katagiri Roshi calls sentient beings that stopped me. There they are — somewhere between birds, humans, and rivers lie the sentient pebbles.

In Maldonado Darwin encountered hundreds of Burrowing Owls, long-legged, round-headed owls, brown with creamy speckles and markings. They are officially crepuscular but are often seen perching or hunting reptiles during the day. In his specimen notes, Darwin recorded the owl's tendency to stand near its burrow and "gaze on you." And surely we cannot truly call ourselves "gazed on" until we have fallen beneath the scrutiny of a Burrowing Owl's round yellow eyes.

BURROWING OWLS

These birds were "excessively numerous," Darwin wrote, and "mentioned by all travelers as one of the most striking features in the ornithology of the Pampas." The scientific name for the bird in Darwin's time, as in ours, is *Athene cunicularia* — *Athene* after the goddess of wisdom, following the deep and cross-cultural belief that owls are, in fact, wise; and *cunicularia*, meaning "burrower" or "miner, after the Latin *cuniculus*, "rabbit." "Wise Rabbit Owl" — a fittingly odd name for such an eccentric little bird. In his notes Darwin's label ventures no further than "owl," evidently the best he could do with this genus that had no European presence. (Actually, Darwin just missed it. The Little Owl, *Athene noctua*, was introduced to Britain in the nineteenth century and eventually became quite common. Darwin would have easily recognized it as a relative because of its morphological similarities, though the Little Owl nests in natural wood cavities and does not burrow.)

In North America, Burrowing Owls typically inhabit prairie dog burrows, and the burrows of other tunneling creatures are often deemed suitable as well — armadillos, kangaroo rats, ground squirrels. The birds are capable of excavating their own holes if none are available, scraping and flinging earth with both their feet and their bills, and it seems that the Florida owls actually prefer to dig for themselves. In the face of human-wrought habitat destruction, many Burrowing Owl populations have been forced to exploit new and less satisfactory sorts of open spaces, including golf courses, airports, and fairgrounds. Their numbers are declining over much of their range.

At the end of an owl's burrow, six or more feet long, is a kind of nest chamber — a small, dark, earthen room, sometimes lined with the sun-dried manure of larger mammals, shredded to a fluff. Here the owls lay their eggs and brood their young. But, unlike most birds, Burrowing Owls stay near their nest sites throughout the nonbreeding season. The burrows give them cover during the night on the open treeless expanses where they live, for, although they give the impression of being on the large side of middle-sized, perhaps because of the length of their legs and the strength of their gaze, Burrowing Owls are really quite small; their bodies are about the size of a robin's. Larger owls such as the ubiquitous Great-horned have no qualms about dining upon these nicely sized birds, sister owls though they may be.

In Maldonado, the Burrowing Owls mainly inhabited the extensive bizcacha rodent colonies. As is the case with most observers, it was the birds' gaze, quiet and direct, that first impressed Darwin. "During the open day, but more especially in the evening," Darwin wrote in the *Ornithological Notes*, "these birds may be seen in every direction, standing, frequently by pairs, on the hillock, by their burrows. Whence they quietly gaze on the passer by; if disturbed, they either enter the hole, or, uttering a shrill harsh cry move with a remarkably undulatory flight to a short distance; whence again they gaze at their pursuer." They stand and stare. All owls do this, but there is something about the Burrowing Owl, standing there on its long legs near its hole, that seems exaggeratedly intentional and insistent. It is difficult to watch them without smiling.

Darwin observed one of the owls kill and carry off a small snake, and he was told that these and other reptiles were the owls' primary prey. He wondered over this: it seemed that for such an enormous colony of owls there were not nearly enough reptiles to go around. Meanwhile, he set some traps for rodents, hoping to study the variety of Maldonado's fauna. The birds and the bizcacha came together for Darwin in a simple but significant observation. "If, by the means of the traps, I had not been aware how wonderfully numerous the smaller rodentia are in these open countries, it would have been an enigma to explain the support of such an infinity of owls."

This is the basic predator-prey relationship, the classic introduction to the food web that all schoolchildren learn and that, as a young intern naturalist, I taught to third-graders at an environmental learning center in Minnesota via lynxes and snowshoe hares. The term *food web*, which would grow out of ecological science, had not been coined in Darwin's time, but here in his owl observations he expresses the most fundamental ecological awareness — the understanding that all beings are sustained in myriad ways by the places in which they live and in ways that are not always self-evident. The secret is under the surface but present and real. After their human watchers are tucked in their beds, Burrowing Owls give up hunting reptiles and turn to nocturnal rodents. Thus an expanse of brown, dry earth and a bit of shrubbery richly cradles hundreds of owls, "an infinity of owls," and all of their odd yellow-eyed owl ways.

*D*arwin worked without benefit of the term *ecology*, which was defined by the German evolutionary thinker Ernst Haeckel in 1869, nearly a decade after the publication of the *Origin of Species,* as the total relationship of an organism to both its organic and inorganic environment. The definition of ecology, now a sturdy branch of the biological sciences in its own right, is often presented even more broadly in modern textbooks. The study of "structure and function in nature," reads one. But most texts simply refer to ecological science as the study of the relationships between organisms and their environment, very much in the manner of Haeckel.

No one with a liberal arts degree escaped the core curriculum without a working definition of the Greek word *oikos*, ecology's apt root, meaning a "dwelling place," a "home," a "household." Ecology as a movement grew up in the United States in the 1960s, when I was a very little girl who knew nothing of such things, and made much of the metaphor, a good and true one, of the earth household. "Leaves begin very gradually to fall," wrote Gary Snyder, the ecology movement's poet doyen, in his early book titled *Earth House Hold*, "what crazy communion of the birds wheeling and circling back and forth in calling flock above the pine and against sunset cold white-and-blue clouds — a bunch of birds being one."

Once he had defined the term *ecology*, Haeckel demonstrated very little ecological insight at all. He worked mainly in systematics and embryology, publishing dense biological texts in mystifying German; these he would dispatch to Darwin, whom he approached on bended knee. Darwin's native inclinations were very different. Even when he was a young man watching birds in South America, it is clear that his observations were shaped by a distinct ecological vision. In his case, it was not theorizing about the natural world that led to a vision of ecological wholeness but the reverse. This sensibility, his unforced awareness of biological interrelatedness would inform his life's thought.

The recognition of continuity between an organism and its biological community came easily and unself-consciously to Darwin, and his natural history notes brim with the moments in which he watched such relationships unfold. The dung of steamer ducks was examined by Darwin and found to "exclusively consist of shell fish obtained from the Kelp, & from the shores at low water." He took note of the ducks' attachment to the near-shore botanical community and tied this further to their habits. "They enjoy but little power of diving," he noted, as they make their lives nibbling invertebrates in the shallows.

In Chiloe Darwin watched, and loved, the hummingbirds, delicate and small, skipping "from side to side, amongst the dripping foliage," and appearing out of character in this "region of endless storms." (He labeled the bird, generically, *Trochilus*. From range and habits, I believe it to be the Green-backed Firecrown, *Sephanoides sephanoides*, a smaller hummingbird well described by its common name.) Darwin had watched other hummingbirds sip nectar from fuschia flowers, even in the midst of snow storms. But in the current season there were scarcely any flowers at all — certainly not enough to support the abundance of hummingbirds. Eventually he discovered that the species "commonly frequents open marshy ground, where the *Bromelias* in patches form dense thickets." The little green birds would "dash" into the thick pointed leaves, searching insects out of their winter quarters, thriving on the protein. In a note that appears to have been added some years later, Darwin determined that the plants were not, in fact, Bromeliads, as they "bear fruit like a pineapple." But his observation remains a fine little revelation in ecological watching, similar to that of the owls and their rodents.

Also in Chiloe, Darwin became quite attached to a bird he called "Myothera," a large suboscine we now commonly iden-

tify as the Black-throated Huet-Huet. The name is ono-
matopoetic, and it was called by English visitors the "barking
bird," and "guid-guid" by the Chilotan natives. It was not the
season for nesting when Darwin visited, but he learned that
the guid-guid builds its nest "among sticks close to the
ground." Similar birds, the Tapacolo and Turco, excavate slender earthen burrows for nesting. In a lovely observation, Darwin wrote, "The nature of the country offers good reason,
why this bird & the Cheucau, build in such a different manner
from the tapacolo & Turco; in these forests, it would be
impossible to make a deep hole, in other than extremely
humid soil." Over and over, Darwin's study reinforces his understanding of the simple but significant continuity between a
creature's habits and its earthen substrate.

Darwin recognized a spiral of relationships, a coil in
which humans were caught up rather than one they observed
from the fictitious ethereal remove recommended by the science of his time. Such relationships revealed themselves as an
organism made its way from its body to its place, from its
place to its sustenance, from its own life to others — an infinity of others.

These connections, based in observation but twined with
intuition, were etched deeply into Darwin's understanding of
the natural world and would surface in all of his future work,
including the *Origin*. Chapter three in particular expresses a
keen ecological grounding. After a succinct and readable exposition of the basis and basics of natural selection theory,
Darwin described, as he labeled it in his subchapter, *The Complex Relations of all Animals and Plants to Each Other in the Struggle for
Existence*. He wrote that the checks and balances governing the
"organic beings which have to struggle together in the same

country" were both complex and "unexpected." He could have been thinking of the Burrowing Owls and the surprise of the abundant rodents. In this chapter, he brought several more detailed observations to bear, all of which portray the delicate, complex, and so often unseen relations between soil, plants, insects, birds, and mammals. His central example highlights a barren heath on a relative's estate, where Darwin had "ample means of investigation." The heath had "never been touched by the hand of man," but a neighboring swath, several hundred acres of exactly the same barren land, had been enclosed twenty-five years previously and planted with Scotch fir.

To Darwin, the change in the native vegetation that resulted from the planting of a single species was "most remarkable." Not only was the proportion of the original plants much changed, but twelve species of plant, not including grasses, thrived in the plantations and could not be found in the heath. "The effect on insects must have been still greater," Darwin wrote excitedly, for there were six insectivorous species of birds found on the plantation that were absent from the heath, and two or three distinct insectivorous birds found on the heath that did not frequent the plantation. "Here we see," Darwin marveled, "how potent has been the effect of the introduction of a single tree, nothing whatever else having been done."

He offered several further examples designed to portray "how plants and animals remote in the scale of nature, are bound together by a web of complex relations." But he concluded that, while the dependency of organic beings upon one another does tend to lie between these beings "remote in the scale of nature," it is nevertheless the struggle between individuals of the same species that will be most intense, "for they frequent the same districts, require the same food, and are exposed to the same dangers."

This insight is an early ecological expression — entirely necessary in Darwin's unfolding vision but requiring a level of struggle and chaos that could never have been neatly contained within the image of nature that prevailed in the eighteenth and nineteenth centuries, both in England and on the Continent, where natural theology proclaimed a world that was both harmonious and benign. According to that deeply entrenched view, the economy of nature sustained a perfect balance in which any "struggle for existence" was blessedly necessary in the preservation of harmony, almost as if those furry little bunnies willingly offered their bellies up to the hawks' pointed talons for the sake of a wider accord. In such a system, at once complex and perfect, nothing was superfluous, nothing out of place. Natural extinction was an impossibility, for any change, the loss of any species, would lead to a deterioration of the whole static construct.

Even today we cling to the vestiges of this myth. So often wild nature is made out to exist in a sustained perfection, where each creature thrives in a state of Elysian adaptedness. But, as Darwin recognized, the natural world is a perfect mess. All species are fluid, pressing toward a perfection that none will achieve. Individuals are beautiful and flawed — hungry, restless, and at home upon a soil that is shifting beneath their feet. In his new and more incisive sense of the struggle for existence, Darwin found a deeper balance; it is this struggle, he believed, that leads to each and every physical or even mental change that is worthy of being called adaptation, and it is here, in this movement, that the spinning, fitful harmony of the natural world is truly unfurled and maintained. It is a sensibility embodied by the hungry little Burrowing Owls keeping watch over their earthen households.

In his essay "A Native Hill," Wendell Berry speaks to the reordering of perspective that occurs when one removes oneself from the usual human confines of house and sheltered garden and walks into the woods. "As I go in under the trees," he writes, "I enter an order that does not exist outside, in the human spaces. I feel my life take its place among the lives — the trees, the annual plants, the animals and birds, the living of all these and the dead." And in the face of this experience he asks a surprising question: "How, having a consciousness, an intelligence, a human spirit — all the vaunted equipment of my race — can I humble myself before a mere piece of earth and speak of myself as its fragment?"

"Its fragment," such an odd phrasing. *Fragment* is a word with a typically negative connotation. A disordered personality is said to be fragmented. A piece of something broken unintentionally from something else is a fragment. I asked my five-year-old daughter, as I am prone to do when seeking a response unhindered by years of overanalysis, what a fragment is. Her answer is usually offered by way of metaphor or example. "A fragment is the arm of a gingerbread person that fell off," Claire told me, and walked away. Yes, that's what I thought. The bit that is left after I've dropped something, something I wish were still whole.

So I pause over Berry's utterance. We might be the arm that fell off, but that is not without its own kind of wholeness, a refreshed understanding of wholeness, grounded in a difficult self-abasement. In recognizing myself as a fragment, I simultaneously accept this earthen dwelling as an element of my origin and of my completion. We are separate individuals, certainly, wonderfully separate, with outlines that are traceable, bodies and minds that are filled with experiences, notions, blood, and bones that are uniquely ours. Yet all of this is wended, surely and deeply, to a wild earth that founds and sustains us, that cradles our past life and proclaims our future. We are fragments — from one angle incomplete, from another entirely whole. We lie in the spaces between these concepts, our feet stepping neatly about the work of our household — kitchen, garden, laundry, child — our jagged backs against the soil, where our edges are smoothed, where all the sentient beings (sea stars, owls, trees, and stones) become our quiet, unseen company.

Most of us live in houses, not beneath the trees, so this serene awareness of ourselves as earthen fragments is necessarily fleeting, and secret. I have noticed that the days I lose this awareness are the very days I am visited by more than my fair share of spiders, dark in the corners of my house, traversing the threshold easily. They lie there, tiny messengers with curled legs, reminders of a wider dwelling.

When I began graduate school in environmental ethics and philosophy, I was particularly interested in questions that had to do with individual wild animals. Although I had strong personal feelings about the issues, my academic interest was not in the usual animal rights questions — animals as food, animals in research, animals as fur coats — but in the idea of animals residing in ecosystems. I had been interning as a raptor rehabilitator, where significant time and resources were spent laboring over the care of relatively few hawks and owls, poor sick birds that would probably die anyway. At first I questioned the rationale for directing our energies this way. Surely the resources, including my fifteen-hour days, would be better spent on conservation of habitat — protecting, on a much larger scale, a far greater number of birds like these as well as countless other creatures and the ecological systems that sustain them.

I fretted over such questions until I was quite sick, actually, walking around with a headache and a nauseous stomach every day. Certainly no one was out patrolling the wilds for sick or injured birds, I reminded myself. At our raptor center, and others like it, nearly all of the birds treated had sustained human-caused injury. Most had been hit by cars, but others had found themselves tangled in barbed wire or had been shot or — even in this time of supposed ecological enlightenment — poisoned as "vermin" predators. Given this reality, it seemed to me that we bore an honest responsibility to these particular individuals; as humans we had harmed them, and as humans we ought to be lifting them up and, if at all possible, returning them to their natural lives.

But I came to see that my thinking on this issue was still oversimplified. Most of the hawks and owls in my care spent the night in their little hospital enclosures, but the intensely sick, or the very young, had to come home with me for overnight care. Other than the Great-horned Owls, who typically smelled like their last meal of striped skunk and so were banished to the laundry room by my housemates, these intensive-care birds slept in small boxes next to my bed, where I could hear them wheezing. For a brief span in my life, I was in the near-constant company of hawks and owls.

Raptors are the very wildest of birds. People dropped off all kinds of birds at the raptor center, and most of them seemed to come to some kind of terms with their temporary life as an enclosed, injured bird. Confined crows looked forward to their vitamin-fortified hamburger meals and called boisterously if we were late with them. Mourning Doves walked quietly up our arms and rested under our hair, close to our necks. Cedar Waxwings seemed positively pleased with their entire captive experience, nibbling blueberries and closing their eyes when their smooth bellies were stroked. Nearly all the birds, if they were not as receptive guests as the waxwings, at least eventually became indifferent to us, paying us little mind as they tossed birdseed about, giving us perfunctory pecks as we caught them up to check a wound or proffer medication.

But the hawks and owls were different. Owls glared intently, each and every time we approached, as if our very presence were the deepest possible affront. Hawks looked through us, as if we were specters, refusing to leave the refuge of their own wildness even to look at us. These birds always carried something of their forests about them, utterly refusing to become captives, even within the smallest enclosures. I found it difficult to invade the wild, psychic lines they drew about themselves, to change a bandage, to shoot them with antibiotics, to lift and lower their sore wings over and over again in a feeble effort to keep scarring muscle tissue supple.

In time, it became clear to me that the idea that we ought to be spending our money on habitat rather than on individuals was based on a separation of creatures and wilderness that was dubious, or even false. These birds, these very individuals, sustain the wilderness from which they have been torn. We cannot consider them separately. It is imperative that we lend our most radical intelligence to the protection of ecological systems, but in the mathematical modeling this nowadays requires, it is likewise important to keep our bearings, to remember, in our references to creatures and species as points of data, the flesh and blood to which we speak. Even our best conservation efforts, which rightly focus on systems rather than individuals, need not remove us from our simultaneous responsibility as earthen fragments, to recall and to respond to the enlivening presence of individual animals within wilderness — a wilderness that cradles all creatures separately, even as it swallows them whole.

Katagiri's very inclusive list of sentient beings, extending even to the stones, begins to make a kind of sense. The stone upon which my hidden oystercatchers stepped was an ocean stone, carpeted by invertebrates, tracked by shorebirds, washed by seawater full of microscopic beings. Its presence cannot be separated, not meaningfully, from this inspirited net of living things. The subdued and basic awareness of the limpet, the bright consciousness of the oystercatcher, the intelligence of the Dall's porpoises that curve, black-and-white, in the near waves — all of this is connected, all is shared. The oystercatcher itself has no being apart from these quiet mussels, this thoughtful stone. Any sentience an organism may lay claim to cannot be counted as its own dear possession. As the stone lends us our footing, we lend back our consciousness in a spirited, inevitable blending.

What would Charles Darwin think of my little reverie here, dipping into transpersonal ecology and the far reaches of the naturalist's faith? I am quite sure this train of thought would seem fuzzy to Darwin, would make him feel like yanking out his hair while proclaiming that his brain had turned to mush. Still, I find my footing in his thought, and in his ways. I recall Walt Whitman's lines "I swear the earth shall surely be complete to him or her who shall be complete! / I swear the earth remains jagged and broken only to him or her who remains jagged and broken!" No matter how curmudgeonly Darwin became (and he became *very* curmudgeonly), he rarely failed to lend his responsive presence to the natural world. His moments there were so often graced, and far beyond the norm. The dim but necessary memory that binds us as fragments to our earthen dwelling rested easily upon his shoulders, and within his watching.

CHAPTER NINE

Wildling in the Galápagos

There is one fact which is extremely singular in the
Natural History of these Islds; it is the tameness of
all the land birds. . . . The Thenca has drank water, out of
the back or shell of a Tortoise, held in my hands, & has so
been lifted from the ground; I have even tried to catch
them by the legs, but failed. In attempting to explain this,
we must remark that no rapacious hawks or quadrupeds
are found here; the only large animal is the harmless Tortoise. Do the birds mistake Man for this huge Reptile?

— *Ornithological Notes*, GALÁPAGOS ISLANDS, 1835

The *Beagle* and its inhabitants traveled as far north as
Lima on the western coast of South America before
sailing east on September 7, 1835, for the Galápagos
Islands. It was calm and gray, and Darwin had mixed feelings.
"In a few days time the *Beagle* will sail for the Galapagos Isds,"
he wrote to Professor Henslow from Lima. "I look forward

with joy & interest to this, both as being somewhat nearer to England, & for the sake of having a good look at an active Volcano. — Although we have seen Lava in abundance, I have never yet beheld the Crater." Darwin's joy was not of the straightforward and bubbling sort but rather hinged on an anticipation that drew deeply from his divided inner world, for as the journey wore on he was increasingly of two minds.

Darwin looked forward to exploring the Galápagos, in spite of reports from earlier explorers regarding the archipelago's desolation and parched wildness. Darwin and the ship's officers passed around the account of Captain George Byron (uncle to Lord Byron, the famous young poet, already dead), who described a volcano that "burns day and night." If anything would whet Darwin's sense of purpose it was this, for in spite of his abiding interest in all things volcanic, he had never actually explored a crater, or *"the* Crater," as he more ominously termed it. This would be his chance.

But even this level of interest was partly drummed up. What filled Darwin's mind as much as any sort of rock was the thought of returning to England. For even though the *Beagle's* planned course included stops in New Zealand, Australia, Africa's Cape of Good Hope, and the Keeling Islands, none of these would entail explorations nearly as long or as involved as those already accomplished in South America. Darwin viewed the Galápagos as the beginning of his journey's end. He couldn't wait to get home.

Some weeks earlier, Darwin wrote a letter to his sister Susan that betrayed this dimension of his mind-set. Besides the obligatory digression asking Dr. Darwin's indulgence with his spending habits, Charles wondered about all that had happened in his absence. He had learned via the lagging naval post that his dear friend and cousin William Darwin Fox had married. "How strange it sounds to hear him talk of 'his dear little wife,'" Darwin wrote, and then added in a catty aside, "Thank providence he did not marry the simple charming Bessy. — I shall be very curious to hear a verdict concerning the merits of the Lady." Darwin's ability to gossip like a schoolgirl is, in my estimation, one of his most redeeming personal qualities. How dull he would have been if he had always been polite and decorous and full of good will. But we need not worry about such things. "How the world goes round," he continued. "Eyton married. I hope he will teach his wife to sit upright." Darwin longed to stick his nose in his friends' personal lives and — it is not too difficult to read between the lines here — to get on with his own, at least as much as he longed to stick his nose in a volcanic crater. But the Galápagos loomed, and Darwin rallied, writing to Caroline, "I am very anxious for the Galápagos Islands, — I think both the Geology & Zoology cannot fail to be very interesting," and also to the freshly married Fox, "I look forward to the Galápagos, with more interest than any other part of the voyage."

Still, as the *Beagle* approached the northwest end of Chatham Island, Darwin scowled. The whole place appeared to be black lava, entirely blanketed with leafless brush and stunted trees and showing few signs of animal life. "The black rocks heated by the rays of the Vertical sun like a stove, give to the air a close & sultry feeling," he recorded in his diary that day, seemingly forgetting his predisposition to any kind of joy whatever. "The plants also smell unpleasantly."

There is no doubt that the Galápagos would become very important both to Darwin and in the development of Darwinism. But the Galápagos figure centrally in various entrenched beliefs within the vast lore of Darwiniana that continue to pervade the general public's mythology of Charles Darwin and his work — this in spite of being popularly debunked in well-respected scholarly works such as Frank J. Sulloway's series of papers for the *Journal of the History of Biology* in the 1980s and equally wonderful popular works such as Jonathan Weiner's *Beak of the Finch* (drawing heavily on Sulloway).

One belief is that Darwin was the first to discover (or notice, or invent) evolution in nature, that he wrote the "Theory of Evolution" in his *Origin of Species*. But, in fact, in addition to making evolutionary ideas more lucid and believable, Darwin's great contribution was his understanding and elucidation of natural selection — the differential reproduction and survival of individuals as the mechanism of evolutionary change. Theories of evolution were already in vogue, particularly Lamarck's notion of evolution via an organism's bodily adaptation to its outward environment and lifestyle. Darwin's eccentric grandfather Erasmus Darwin was a self-styled Lamarckian who wrote reams on the subject, often in poetic verse, including an enormous book titled *Zoomania*, popular in his circle. Philosophers at least as far back as Aristotle questioned the stability of species, though Aristotle's version bordered more on actual transmogrification than on evolution. He believed, for example, that the migratory redstarts that disappeared from the Greek Isles each fall actually turned into the robins that appeared for the winter, and then turned back again into redstarts. (This was fairly decent natural history for its time, based just as surely in observation of living things as Darwin's would be.) Darwin did not invent or discover the fact that species evolve over time into new and different species. He simply became convinced of evolution's truth and then worked the rest of his life to explore and explain its ways.

Another misconception is that Darwin became converted to the truth of evolution while in the Galápagos, studying the famous finches with their adaptively radiated bill sizes and shapes. According to the most extreme version of the myth, Darwin observed the finches carefully while exploring the islands, noting with some excitement the differentiation among the bills of these very similar species and, further, that finches on differing islands were adapted to their circumstances in particular ways. As a result, he came to understand, quite swiftly and insightfully, that geographic isolation and natural selection led, over evolutionary time, to these differences. The *Origin of Species*, and all of Darwin's future work, the myth continues, would draw on this understanding.

In fact, Darwin's grasp of finch biology was murky at best. He misidentified many of the birds, failing not only to differentiate between similar finch species, but also to recognize many of them as finches at all. Galápagos finches were labeled variously by Darwin as finches, warblers, and wrens. Darwin did not pretend otherwise. "Amongst the species of this family," he wrote in the *Ornithological Notes*, "there reigns (to me) an inexplicable confusion." He utterly missed the salient point that geographic separation has a part to play in species differentiation and did not even bother to label his species according to the island from which they came. He assumed that all the birds in such a small archipelago, while different from the mainland avifauna, were similar between islands. He did quite desperately attempt to label species by their island of origin after he returned to England, when an intimation of the potential import of such details began to dawn. He drew on the collections of other voyage members, particularly FitzRoy's and Sims Covington's, who had labeled their birds more stringently, to surmise the probable origins of his own specimens.

FOUR SPECIES OF GALAPAGOS FINCH, DEMONSTRATING BILL VARIATION

This post hoc labeling has been responsible for much of the confusion in scholarship regarding Darwin in the Galápagos — earlier historians assumed that since many of the existing specimens are labeled as island specific, Darwin must have originally labeled them this way in situ. But the existing labels, though old, are not Darwin's own. His field labels were replaced with more permanent and legible labels by collection curators and often according to Darwin's new guesses about the island from which each bird had come, which may or may not have been correct. Everything was a terrible muddle for the longest time, and only recently have scholars begun to set things right.

The notion that Darwin had any decent insight at all regarding the finches while in the archipelago is further dispelled by the observations that he did make. Darwin found the fact that some of the birds were black to be quite interesting and assumed it was a rarer color, or perhaps indicated a separate species. While he had learned through his own earlier observations that plumage often changes as an individual bird matures and that there are juvenile and adult plumages (though this terminology had not been codified in his time, as it is today), Darwin did not seem to consider such plumage development in the case of the Galápagos birds. He wrote a long paragraph in the *Ornithological Notes* about the apparent mystery of black finches and just one short note on their bills:

"A gradation in form of the bill appears to me to exist." He did not apply himself to sorting this variation out, so, as the differentiation in color distracted him, the differentiation in bill size and shape eluded him. Without John Gould, Darwin would never have recognized the Galápagos finches' taxonomic unity. Yet the Galápagos myth gained incredible influence during the twentieth century. As Sulloway wrote in his seminal paper, "Darwin and His Finches: The Evolution of a Legend,"

> In spite of the legend's manifest contradictions with historical fact, it successfully holds sway today in the major textbooks of biology and ornithology, and is frequently encountered as well in the historical literature on Darwin. It has become, in fact, one of the most widely circulated legends in the history of the life sciences, ranking with the famous stories of Newton and the apple and of Galileo's experiments at the Leaning Tower of Pisa, as a classic textbook account of the origins of modern science.

Sulloway was writing two decades ago, and though the myth has been dismantled little by little since that time, elements of it still persist. Even most teachers and students of biology and its related disciplines would be surprised to learn that the Galápagos finches are not specifically mentioned in Darwin's *Origin of Species*, not even once.

The phrase "Darwin's finches," now so ingrained in the mythology of science, seems to have first appeared around the turn of the twentieth century, but it was brought deeply into modern parlance by the biologist David Lack, who studied the finches in the forties and published his beautiful book *Darwin's Finches* in 1947. Here, Lack made one of the most eloquent statements to date of the evolutionary synthesis that marked the thirties and forties, exploring the interactions between genetic variation, geographic isolation, and classical natural selection through the thirteen species of Galápagos finches. He used an epigraph from the *Origin of Species* to begin each chapter and sprinkled the entire book with Darwinian lore, some of it (such as the notion that Darwin had actually separated his avian specimens according to island) unintentionally erroneous. Lack's book was widely read and served to cement the conflation of Darwin's actual experience in the Galápagos with birds Darwin had never identified and thoughts he could not have had, as much of the science Lack discussed was not yet known by the time of Darwin's death. Lack surely had no diabolical intentions — he himself was simply caught up in the legend and was perhaps inspired by a sense of his own work's living squarely in the Darwinian lineage, as it surely does. But Darwin arrived upon and left the shores of the Galápagos a creationist — a malleable creationist, but a creationist nonetheless. And while these various intertwined myths contain seeds of truth, none of them is quite right, and none of them is nearly as wonderful as the truth.

It is well known that most creatures inhabiting the Galápagos archipelago, having evolved in the absence of mammalian predators, are rather tame. Darwin, with his secret Dr. Doolittle–style ways, was used to drawing nearer to animals than most of us plodding humans are able, but even for Darwin, the Galápagos became a veritable playground. He was sure of himself as a naturalist by this time, and it seems he approached his record keeping with less to prove. He turned a little giddy and loose in his observations, and his diary became, quite suddenly, a great deal of fun to read and, I presume, to write. He may not have been at his most intellectually acute, but he was interested, thoughtful, and attentive to the creatures in whose midst he spent his strange island days. Darwin slipped with ease into the wildness of the place, and there is much to be gathered from Darwin's time there, *while* he was there, not in terms of what it would inspire in his later thinking. In the Galápagos, Darwin stirred up his increasingly odd inner brew. Aware of the nearness of his journey's close, he was, more than ever, concerned about the scientific life that awaited him in London. At the same time, the Galápagos revealed him to be, more than at any other time or place on the journey, at ease, at home, and at play in the remote wild.

Exploring Chatham Island on September 17, Darwin first observed the ubiquitous marine iguanas, and he appraised them most unsympathetically: "The black Lava rocks on the beach are frequented by large (2–3 ft) most disgusting, clumsy Lizards. They are as black as the porous rocks over which they crawl & seek their prey from the Sea. — Somebody calls them 'imps of darkness.'" Later, Darwin would borrow the ship surgeon's scalpel to "open" the iguanas and discover to his surprise that the "prey" they sought in the sea was entirely botanical. They ate seaweed. He watched them for hours, languidly draping themselves over the rock. "They assuredly well become the land they inhabit," he wrote, in a kind of backhanded compliment.

Regarding the archipelago's reptiles, Darwin much preferred the tortoises. Walking with Covington on Albemarle (now Isabela) Island a few days after meeting the land iguanas, he encountered two "very large Tortoises." They behaved well. "One was eating a Cactus & then quietly walked away. — the other gave a deep & loud hiss & then drew back his head." Surrounded by the black lava, the odd stunted trees, and giant cacti, the tortoises "appeared most old-fashioned antedilu-vian animals; or rather inhabitants of some other planet." Darwin rode slowly about on a tortoise's arched carapace; the large old males carried him easily, and drew him into their in-terplanetary landscape.

"The tortoise when it can procure it, drinks great quanti-ties of water," Darwin wrote in his diary about one of his fa-vorite tortoise activities. They would "swarm" about a spring, coming and going from the water pool, lines of tortoises walking resolutely in two directions. "The effect is very comi-cal in seeing these huge creatures with outstretched neck so deliberately pacing onwards." When the tortoises arrived at the spring, they would bury their heads, plunging even their eyeballs beneath the water, and "greedily suck in great mouth-fuls, *quite regardless* of lookers on." I love this last — as if an English tortoise might have behaved more daintily in com-pany.

Darwin was told that the old tortoises occasionally die by falling over cliffs, but that "the inhabitants have never found one dead from Natural causes." Tortoises seemed to live on and on, growing very old, their dying unknown. How odd. Surely this reinforced Darwin's sense of the strange enchant-edness of this land — a place with sighing reptiles that live outside of time.

Darwin found the birdlife of the archipelago to be very pleasing, partly because he could observe it so closely. He marveled as a hawk he approached refused to move from its branch until he eased it off with the tip of his gun. He was charmed as a mockingbird sipped water from a small tortoise-carapace cup he held in his hands, and for once he managed to delight in the moment rather than thwacking the bird on the head and dropping it into his field bag. Darwin had come to the Galápagos to walk upon "the Crater," and instead he swam with iguanas, sunned himself on the back of a tortoise, and pranced about with small brown birds perched on the brim of his hat. He loved the Galápagos in ways that he never expected, smelly plants and all.

His notes on the birds are thoughtful and attentive, though certainly not his most brilliant. He attempted, with two ship's officers, to collect an example of all the land birds, and believed himself to have succeeded quite well. "There are no Hawks besides the Caracara," he wrote, which is somewhat odd because the only hawk (besides the occasional Osprey and Peregrine Falcon, neither of which is endemic or noted by Darwin) is the Galápagos Hawk, actually an endemic buteo in the same genus as our ubiquitous Red-tailed Hawk and the smaller Broad-winged Hawk. Darwin usually did quite well at identifying raptors. But the Galápagos Hawk is confusing in that it is very dark in color, long and slender, and eats carrion so voraciously that its crop becomes distended, much like the caracaras Darwin had observed on the South American main-land. Still, there is a difference in skeletal structure, evident in the shape of the head, and also a particular silhouette to the wing, evident in the birds' manner of soaring, that Darwin might have been expected to discern in one of his more obser-vant moments. In the *Zoology*, Gould acknowledged the hawk's affinity with the buteos, or "buzzards," as they are still known in England. But in that same volume Darwin rather defen-sively pointed out that the bird would have to be considered a caracara "if a principle of classification founded on habits alone were admissible."

Darwin's favorite birds on the archipelago seem to have been the mockingbirds. They reminded him of the mockingbirds he had observed on the South American mainland, the "Thenca" of Chile (now called the Chilean Mockingbird), the "Calandria" of La Plata (Chalk-browed Mockingbird), and another similar bird in Patagonia, one that Darwin didn't know whether or not to label as a separate species from the others but that Gould rightly determined to be unique, the Patagonian Mockingbird. It is clear from his

CHATHAM ISLAND
MOCKINGBIRD

notes that Darwin thoroughly enjoyed these birds' brazen habits. "They are tame & bold," he wrote, and "constantly frequent in numbers the country houses to pick the meat; which is hung up on the posts or walls. — if any other small bird joins in the feast, the Calandria directly chases him away." The Patagonian bird, he noted, was "rather wilder; it there commonly haunts the valleys clothed with spiny bushes on the higher twigs of which it takes its stand." The mockingbirds' ability as songsters particularly impressed Darwin, and he heaped praise upon them. "The song is remarkable, as being far superior to that of any other South American bird; indeed I have not heard any other bird, which properly perches itself to give continuous music."

Regarding the Galápagos birds, Darwin wondered at first whether they were the same species he had observed in Chile. "In their habits I cannot point out a single difference," he wrote in the *Notes*. "They are lively, inquisitive, active, *run* fast, frequent houses to pick the meat of the Tortoise, which is hung up, — sing tolerably well." They get into all the mischief that good mockingbirds are prone to. And, of course, they are

"*very* tame, a character in common with the other birds." But
then he wondered, questioning, "I *imagined* however its note or
cry was rather different from the Thenca of Chile?" Darwin
knew that each avian species possesses a distinctive song and
call, and his travels would have attuned his ears to the variation in regional dialects that bird species often display. This
seemed to be another "cry" altogether. Why would this be?
How could a bird so similar be a different bird?

Darwin grasped the fact that this was a deeply meaningful
question, even if he was not quite ready to admit just how
deeply meaningful it really was. The Galápagos mockingbirds, like much of the flora and fauna on the archipelago, appeared to be affiliated with South America, making the
coastal mainland the "centre of creation" for the islands, the
place from which the species that populate the islands originated. The fact that the Galápagos species looked somewhat
different from the mainland species was one that lay within
the parameters of special creation — even animals of the
same species might vary somewhat in order to befit the particular place they lived. But what if this mockingbird really
was a completely different species? That would make no sense
at all.

According to the tenets of special creation, species were
designed to suit the place where they lived, so the presence of
the endemic marine and land "lizards" on the Galápagos was
not a stumper. They blended so well into the landscape, as
Darwin noticed, they looked like they might melt into it at
any second. Clearly they were made for the place. But these
birds were different. The Galápagos, as Darwin noticed over
and over, was entirely dissimilar from the mainland, its own
little world, enchanted, apart. If the Creator was populating it
with new species, they should be very different from those of

Ecuador and Peru. If the Creator was using South America as the Galápagos' center of creation, then the species should be the same.

This was Darwin's earliest declaration regarding his increasingly uneasy sense that something was fishy about special creation. Clearly such a statement would fuel the myth of Darwin's Galápagos conversion to an evolutionary mind-set. But Darwin did not pen these notes while in the Galápagos. Frank Sulloway, paying Sherlock Holmesian attention to such matters as Darwin's evolving spelling habits and the watermarks on his notepaper, dates this passage to August of 1836. That would be some nine months after Darwin left the Galápagos, when, during the return voyage, he was copying the ornithological notes from his general specimen notebook, where there is no such musing. While most of the *Ornithological Notes* were copied almost exactly from the specimen book, this is one of the instances in which Darwin's past observations and current thoughts became wrapped up together in a new synthesis. This was still the work of a questing mind, not of a committed evolutionist, and Darwin would backpedal twenty miles on this notion before he would go one forward. He had, as yet, no conscious inkling of how natural selection might work, and when he returned to London he would still call himself a creationist. But he was a questioning creationist by that time, to be sure, one who hid in the back of his mind a place where the fact of species transmutation had taken root. Passages such as this one show us that the Galápagos, while it was no Archimedes' bathtub, sent Darwin home with a mind full of bits to be mined and pondered.

Maybe the myth should have unfolded as we like to imagine it. Or at least maybe we would all be more pleased and inspired if it had. If the young Darwin, trodding these bleak islandscapes with nothing but a well-worn copy of Milton's *Paradise Lost* in his pocket, alone, monkish, and brilliant, had seen various finches — one with a large bill cracking thick seed pods, one with a small bill picking dainty ones, and others in between — watched them in a long, suspended silence, and then uttered, "My God, I see." There is expectation, and desire, and even comfort in the imagining of such moments. We welcome a belief in the human capacity for shocking, instantaneous brilliance: Isaac Newton's apple, Darwin's finches, Benjamin Franklin's kite. We long, ourselves, for moments of raw insight, and such tales fit them in our minds, making them feel palpable and within reach. But the truth about the creative mind and the nature of scientific insight is still lovelier. Like earthly life, insights evolve. They may appear to make a sudden leap, but these leaps are built on the foundation of work, knowledge, and vision.

Darwin returned home to London. He read Plato, Wordsworth, and even more Milton. He consulted with Gould, with Lyell, with Henslow, with farmers, and pigeon fanciers, and economists. He began a voluminous correspondence (we have fourteen thousand extant letters), asking everyone he could think of anything he could think of that bore on the problem that formed in his mind with increasing clarity. He mingled these eclectic studies with the tangle of experiences and observations from the *Beagle* journey that now lived within him, within the naturalist he had become. He engaged in a process of creativity that was vast and extended and weblike, and one that bore absolutely no resemblence to a logical chain of events where this happened, then this happened, then this happened, and then Darwin said "Aha!" As Stephen Jay Gould put it in his opus *The Structure of Evolutionary Theory*,

"Darwin's greatest intellectual strength lay in his ability to forge connections and perceive webs of implication (that more conventional thinking in linear order might miss)."

I have to admit to harboring an irrational little bias against the Galápagos birds and all the attention that has been lavished upon them over the decades in any discussion of things Darwinian. After all, the birds of the Galápagos were only a few of the birds that Darwin experienced and learned from on his five-year journey. It was during his years of circumnavigating southern South America that Darwin developed his own unique sensitivity to avian beings. If he had not watched the Chilean Thenca so well and so dearly, he would have had no grounds for his significant and lovely questions regarding the Galápagos mockingbirds. I find the whole South American season to be so inspirational, watching Darwin develop into a decent, and then a practiced, and then a beautiful, watcher, and yet all of the non-Galápagos birds are rarely explored in the literature. In bringing many of these birds to light, I hoped to recover their significance for Darwin in relation to the more famous Galápagos birds. I thought I might even set things further aright by slighting the Galápagos finches and mockingbirds altogether — not writing about them at all, paying all my attention to the quiet and less assuming birds that few ever connect with Darwin. But I couldn't quite bring myself to do it.

For one thing, although Darwin grossly fumbled his handling of the finches while on the archipelago, and while his facts regarding their distribution within the islands were never straight enough to allow for inclusion in the *Origin*, John Gould's deft identification of the birds as thirteen different species of the same genus was among the single most important events that, upon Darwin's return to London, began to both give shape to his evolutionary understanding in general and to sow the seed of natural selection theory in particular. (At a meeting of the Zoological Society in 1837, Gould unveiled his name for the new genus — *Geospiza*, the name we use today — and he also recognized subgenera that have since fallen by the taxonomic wayside. Modern ornithologists recognize fourteen species of *Geospizinae*, a number very close to Gould's thirteen, though the arrangement is quite different.) If the Galápagos birds ever visited Darwin in a moment of raw, pure, gritty insight, it must have been then, when the meaning of Gould's identification of the finches and mockingbirds began to settle in. But it seems to me there is another reason that Darwin would eventually, in his autobiography, call the Galápagos the "origin of all my views" — a reason that Darwin himself did not articulate, a reason less tangible but just as significant as his bag of misidentified finch bodies.

The Galápagos archipelago was the most remote, removed, barren, alien, completely wild place that Darwin visited, and yet it was there that he experienced his most radical intimacy with wild beings. Because of the extreme tameness of the birds and other creatures, the Galápagos offered Darwin a closeness to animals that surpassed anything he had yet known. The place was fantastic and wholly alien to him, but instead of responding as a stranger in a strange land, Darwin appears to have made himself very much at home. I doubt that this was a conscious decision on Darwin's part, an attempt to feel such normalcy in the close presence of wild things; instead, I envision him, after his long immersion in a wilderness of both mind and place, falling into this new sense of residence with seeming ease, and with delight.

It may have been this sensibility, as much as anything, that allowed the proposal of Darwin's eventual theories. Darwin was about to return to the strict and tidy norms of early Victorian society, but he carried with him the remembrance of this remote and wild place that was entirely real and true, a place peopled by immortal tortoises, a place where he had *lived*, a place that could never be folded neatly enough to lie within the philosophy and theology of his time. In this light, the Galápagos facilitated Darwin's break from the normative thinking of his day — a break that was a necessary step in the evolution of his revolutionary philosophy with all of its radical implications, for which so much of human culture, even today, seems not quite prepared. Darwin's vision came to reveal a natural order that refuses to mark humans as separate or exceptional or beyond the reach of wildness. His is a complete reversal of the accepted universal order; a movement from the Platonic notion of the perfect archetype to the Darwinian notion of variable individuals; the inevitable recognition that the literal conception of a personified Creator-God is primitive, simplistic, and not particularly relevant; the further insistence that any sort of God left would be one that refuses to impart order and harmony upon a natural world filled with contingency and chaos. No wonder poor Darwin, whom self-help experts today would surely label a "people-pleaser," quaked fearfully in his shoes for twenty years before making his ideas public, and even then had to be nearly forced to do it by the young Alfred Russel Wallace, who was seeking publication for his own shockingly similar ideas on evolution by natural selection. No wonder Charles labeled his first notebook on the subject, begun just as he returned from the *Beagle* voyage, "Nothing for Any Purpose" and kept it an absolute secret, making the description of the Galápagos that he published in the *Journal* an apt one to depict the place his own wild mind had become — an enchanted place, "a world apart."

The word *wild*, and its various incarnations, keeps springing up in relation to my interpretation of Darwin and his thoughts. I am reminded of Thoreau's famous quotation, "In wilderness is the preservation of the world," which so often appears on sunny nature posters and calendars. Certainly it is a lovely thought, but, as Jack Turner points out in his challenging book *The Abstract Wild*, it is not what Thoreau wrote. He wrote, "In *wildness* is the preservation of the world." Turner suggests that wilderness is an important but on the whole lighter concept, appealing to our aesthetic sense of what a wilderness embodies (something very similar to what is on the posters where the misquotation typically appears) — a green and leafy place, slightly out of focus, a place to lift our spirits as we watch the sunlight sparkle upon the babbling stream. But Thoreau never rested in the aesthetic dimension. What could he have meant by *wildness?* The word *wild* embodies true wilderness, certainly, but it is still more complex; it is a place, a quality, a concept, all twined to become a new sensibility, a way of being.

In *Origins, A Short Etymological Dictionary of Modern English,* Eric Partridge listed the various words that relate to *wild* in some way, and one of them particularly struck me: *wildling.* The suffix -*ling*, "a person, animal, thing that belongs to something," in this case, the *wild* — the uncivilized, the uncultivated, and hence "savage." The link between the wild and the savage runs deep — it is little wonder we try to soften Thoreau's wildness. *Wildling* refers particularly to "a wild plant transplanted to a cultivated spot," as Darwin would be in London, as humans are at their most wonderful — living in culture, yet not contained by the parameters of civilization, not completely sealed within a sense of entitled human separateness. To live in culture well, and joyfully, to live within Thoreau's vision (and Darwin's), means that we live with a vision of the truth and presence of wildness on earth. We allow this vision to penetrate and inform our lives. We yield to its direction of our actions insofar as they affect all of life — human and nonhuman. We allow this vision to delight us, to inspire us, to change us.

Darwin's conversion as a naturalist, the shift that began so dramatically in Brazil and continued more subtly throughout his journey, came to a kind of ripening in the Galápagos, marked by such wild intimacy. And as the Galápagos was, in Darwin's mind, the beginning of his return journey, this sensibility became strongly entangled with his thinking about his future life, a future that was suddenly approaching. For Darwin, the psychological sense of these last weeks as a return journey ran deep. He was going home, and he meant to step upon his home shores as a kind of new creation. His apprenticeship was, in his mind, over; he was ready to present himself, finally, as a "finished naturalist." Still stashed away on the *Beagle*, Darwin was beginning his public scientific life. In the preparation, and in the waiting, his homesickness grew almost unbearable. Still, he studied hard, organized his collections, and applied himself rigorously to the copying of his hugely daunting collection of notes. It was interactive work, the ideas in his mind, jostled forward by his experiences, mingling with his notes on paper, the specimens in jars, the drawers full of feathers, the tangible fruits of his journey.

When Darwin left the Galápagos, he certainly didn't know that the gray bearded face of his mature self would be pasted lavishly in every biology text published in the entire next millennium, but he did know something: He really was a scientist. It would be within the scientific world that he would make his life, not as a happy bug-collecting clergyman. In spite of the fact that there was more exploring to do — dips into Tahiti, New Zealand, the southern tip of Africa and Australia — Darwin's naturalizing diminished. He still kept his diary, but it reflects a more general level of daily experience — dinners and meetings with local people, a little sightseeing, much less natural history, and more internal work. The *Ornithological Notes* end with the Galápagos birds.

CHAPTER TEN

Living Science, Dead Birds

With its tail erect, & stilt-like legs, every now & then
[the Turca] may be seen popping from one bush to
another, with uncommon celerity. It really requires little
imagination to believe the bird is ashamed of itself & is
aware of its most ridiculous figure. — An ornithologist,
on first seeing it, would exclaim, "a vilely stuffed specimen
has escaped from a museum & has come to life again"!

— *Ornithological Notes*, VALPARAISO, 1834

The *Beagle* finally docked at Falmouth on October 2,
1836. Darwin raced home, and even with the coach's
horses running at a full gallop, the journey to
Shrewsbury took two full days. He arrived in the dark, when
the entire household was asleep. Hardly able to contain himself, Darwin politely slipped into his old room without waking a soul and fell into an exhausted sleep. The unfettered
ecstasy over his appearance at breakfast the next morning met

all his dearest expectations. He was hugged. He was kissed. He was fawned upon by loving sisters. He was too thin! He had grown tall! He was much altered! No, no, he was the same old Charles. Darwin gave himself over to the delight of it all, giddy, happy, home at last.

But the cheerful simplicity of the moment did not last long, at least not for Darwin himself. The profusion of baggage he carried kept creeping into the forefront of his mind and imagination. He'd left the *Beagle* with a deep, ripe longing to be home, yes, but also with a veritable hoard of scientific beginnings. There was, of course, his diary, all 770 pages of it. There were his notebooks — 1,383 pages of geology, 368 pages of zoology. There were his specimens — 1,529 creatures (or at least bits of their innards) in spirits, 3,709 stuffed birds, animal skins, bones, and dried pieces of this and that. There were crates of fossils, birds, corals, and rocks that he had shipped ahead from the *Beagle*. There was the half-eaten Avestruz Petise. There was the baby Galápagos tortoise that he had carried home alive, and it had grown more than two inches already.

All of these material trappings proclaimed the outer manifestation of a deeper dissonance. For in spite of various exclamations to the contrary, he was anything but the same old Charles. This was a young man who had lived richly at the far reaches for five years. He had experienced true wilderness. He had witnessed, and allowed himself to be deeply affected by, the shocking state in which humans might live — both as slaves and as "savages." He had felt the very earth quaver beneath his feet in a major earthquake. He had been sick so many times that it was a wonder he had made it home at all. He had, FitzRoy's solicitous ministrations notwithstanding, carried himself time and again by nothing but his scrappiness and his wits. He had thoroughly remade his life. He had had thousands of experiences, all of them now whirling in his young brain — moments unknown and untouchable to all but himself. He was a complete mess of thoughts, and he quickly became aware of the fact that he was quite alone with them. He knew things to be true about himself that no other person knew, or, he was sure, could have understood. Darwin could not have helped but feel an acute alienation.

He began to beg off social visits, ones that he had longed for just months ago. He even avoided his old kindred spirit, Fox, who was now married and had a baby to show off. Darwin was a tormented scientist in the making; Fox was a father and a clergyman, living the life that might have been Darwin's own. No, that was just too strange. Darwin would write a warm letter to Fox and visit later.

Darwin's withdrawal into this inner turmoil did not go unnoticed, especially among his closest intimates. "He was shy," Elizabeth Wedgwood noted after their first meeting since his return, "we could not get on." Rather than work on his social graces, Darwin chose to retreat more and more fully into the scientific persona he hoped to build for himself. He grew anxious to get at his specimens.

After just two weeks at home, Darwin broke loose from his sisters, rushed off to London, took rooms in Cambridge, and began to gad manically about, visiting Henslow, joining natural history societies, rushing between institutions, and seeking expert naturalists to work on the sorting of his specimens. Henslow had seen to it that several of Darwin's letters and findings had been published while he was still journeying, so his name was already somewhat known. His large fossils of presumably extinct mammals were of particular interest, and gained him some recognition, which made it a little easier to get the attention of busy naturalists who might further his work. After some debate about the best and most appropriate institution for his specimens, he deposited the vast bulk of them with the Zoological Society of London. He was eager to have his treasures scrutinized by the gaze of experts, and he engaged one naturalist to work on the reptiles, another the insects, and the superintendent of the Zoological Society's museum to do most of the rest. John Gould would examine the birds.

Gould himself was an odd egg. Unlike Darwin, Gould actually had to work for a living, and at twenty-three he had begun a career as a taxidermist, rising through the ranks of the Zoological Society and eventually becoming the superintendent of its ornithological collections. He was one of the first professionals in the emerging ornithological discipline, and few could rival his understanding of avian systematics, which appears to have come to him not just through the examination and classification of thousands of specimens but also through a natural gift. Gould could take in the morphological characteristics of a bird, uncommon details that would have been passed over by other eyes, even experienced ones. He could fit these into a vast natural system that, though he referred to texts and collections as he made classifications, he seemed to carry mainly in his brain.

Gould was also a talented and increasingly well-known artist. He sketched birds, sometimes from life, often from mounted specimens arranged in some supposedly lifelike pose (and in this he was far more successful than John James Audubon, who wired his specimen-models into oddly contorted postures). John Gould painted watercolors from his sketches, and his wife, Elizabeth, would transfer the images to the lithographer's stone, enhancing the work in both detail and artistry. Theirs was truly an artistic partnership, though John Gould alone most often signed the prints.

The Goulds' five-volume, extravagantly illustrated *Birds of Europe* sold so well that the pair was able to finance a trip to Australia, where they would write their most famous work, *The Birds of Australia*. The trip left Darwin feeling very much abandoned, for Gould departed with the *Beagle* specimens unfinished, having puzzled out the identities of the rarest and most interesting birds and leaving the rest. Darwin managed to secure George Robert Gray to finish the task, a thoroughly competent but comparatively lackluster replacement.

For all his personal quirks, Gould was brilliant, and Darwin was fortunate to have had him. He distinguished subtle differences that identify two similar specimens as distinct species, where Darwin had thought they were only varieties. In a time when stuffed curio birds were the norm — both for study and for classification — Gould impressed and surprised Darwin by relying on his notes regarding birds' behaviors and habitats to make species designations. Darwin stood dumbfounded as Gould sifted through the bag of supposed finches, wrens, and warblers from the Galápagos and parceled them out into thirteen separate species of a new genus of finch that he named *Geospizinae*. Gould carefully described Darwin's decrepit rhea specimens as two different species, inspiring Darwin's best thoughts on geographic distribution and the making of species. On so many counts Gould did not merely describe and name stuffed bird skins. He educated Darwin and pushed his mind forward. Darwin watched over Gould's shoulder, a rapt pupil.

It is likely that neither Gould nor Darwin realized the extent to which they were at a crossroads in the history of ornithology as a discipline. During the eighteenth century the popular *cabinet d'histoire naturelle* had proliferated throughout France, and then all of Europe. Small private collections of stuffed birds, eggs, shells, insects, and the like were pursued, purchased for astonishing sums, and displayed by the aristocracy. Public "museums" consisted largely of decorative dioramas featuring exotic stuffed birds, eggs, and an assortment of plants arranged in scenes purported to be "naturalistic" (despite the fact that species from vastly differing parts of the globe were often displayed side by side, as if they had fledged from the very same nest). A modest fee was typically charged for admission, and there was little pretense to educating the visitor. Such museums were simply glorified curio cabinets, the sort found in aristocratic parlors but on a larger scale.

In the nineteenth century the situation started to change. Ornithological study began to shift more obviously from the salon to the museum, and these museums — provincial, private, or associated with universities — became more deserving of the name, directing significant resources toward rounding out collections, preserving specimens, proper labeling, and enhancing educational value. Zoological societies often maintained their own ever-growing collections, which they viewed with an increasingly scientific eye. Some even employed a curator. By the end of the nineteenth century, the institution of the natural history museum had truly come into its own, but even nearing midcentury, when Gould was teasing out Darwin's collection, the British Natural History Museum, the Zoological Society of London, universities such as Cambridge, and numerous smaller provincial museums, many with excellent collections of local species, were moving the study of ornithology toward a serious respectability.

Perhaps even more than the proliferation of natural history museums, innovations in the printing industry buoyed ornithological interest and study in both academia and the upper middle class. Paper began to be manufactured in rolls on a machine, printing presses started running more efficiently on steam, movable type gave way to stereotype plates, and rich Morroccan leather could be replaced by more affordable cloth covers in all sorts of lovely colors. The development of lithographic techniques was particularly significant for students of ornithology, affording the opportunity for more lifelike detail, depth, and accuracy — in part because artist-naturalists could render their work directly onto the stone themselves, rather than relying on engravers who often cared little for matters as trifling as ornithological accuracy. Most important, lithographs were far less costly than either woodcuts or engravings. With this newfound efficiency in printing and illustration came a sudden proliferation of beautiful, high-quality, and affordable books about birds. These did not replace the expensive subscription volumes of the sort produced by Audubon and Gould, but they supplemented them for a wider and increasingly diverse audience. These developments coalesced into a focus among the ornithologically inclined on increased specialization and higher standards.

In spite of this progress, all manner of troubles beset the nascent discipline. Most birds of the world had still not even been observed or collected by the few Europeans and Americans involved in such studies, and the bird life of certain areas, such as China and Madagascar, remained entirely unknown. Rules for scientifically naming birds had yet to be clarified, and Linnaeus's famous *Systemae Naturae* was not nearly complex enough to accommodate newer discoveries about avian relationships. Communication between scientists in different countries (let alone continents) was dismal, so most known birds had been "officially" described several times and called by as many differing names. Rules for determining precedence in naming had not yet been codified. The roles of habitat, behavior, and other aspects of avian natural history were underappreciated. The differences between species and varieties were not commonly understood, and collections of stuffed birds that might have been used for reference were turning into dusty, worm-riddled heaps, as developments in the art of preservation were long in coming.

It was clear that if ornithology was going to be taken seriously, some universal classification standard had to be implemented. In the nineteenth century, work in species recognition, classification, taxonomy, and naming depended directly upon specimen collections, and the lack of adequate preservation techniques proved a formidable barrier to forward movement. It was impossible to maintain a serious working collection when the majority of specimens were being nibbled away by unseen pests. Curators tried everything; birds were baked, scraped, soaped, and dipped in all manner of toxic potions, and still they crumbled. As the birds' study skins rotted away, taxonomists were left relying disproportionately upon similarities and differences in bills and feet — the parts of a bird most impervious to pests — in making classifications.

While at Edinburgh University and before he'd ever heard of the *Beagle*, Darwin had engaged John Edmonstone, a freed slave who prepared specimens for the university museum, to give him lessons in the methods of taxidermy. The men met for two months, one hour each day, and Darwin proved an interested and reliable, though rather average, pupil. Edmonstone's lessons would serve him in unforeseen ways. Darwin's theories might have been much longer in coming, had they come at all, without his ability to return with properly prepared (if poorly labeled) avian specimens. Edmonstone taught Darwin to treat his specimens with arsenic — the breakthrough, state-of-the-art preservative that finally transformed nineteenth-century collections.

Because of its toxicity, arsenic is no longer in popular use (some historians believe that Darwin's later health problems were caused, in part, by claustrophobic exposure to arsenic and formaldehyde as he worked on his specimens in the cramped *Beagle* quarters), but because of their age, the bulk of specimens in most collections have been arsenic-treated. Modern curators typically use no chemical preservatives at all, relying instead upon proper storage, with study skins arranged neatly in acid-free cardboard boxes and then placed in special airtight cabinets. At the first sign of insect incursion, entire collections are plopped into the deep freeze. Still, curators live in a constant state of anxiety over the condition of their collections, and several have confessed to me that they wish arsenic were still a viable option. These days even mothballs are frowned upon, their use protested by graduate students with chemical sensitivities.

But besides the arsenic, techniques of avian specimen preparation have changed little since Darwin's time. Beginning with a bird in tolerably good shape, either freshly dead or recently thawed, the preparer, using a scalpel, begins by making a careful vertical incision along the bird's ventral surface, from the neck to the vent. Care is taken to avoid opening the abdomen, as the flow of juices could soil the plumage. Powder is often applied to the preparer's hands and sprinkled into the bird throughout the process, drying any wetness that could cause feather damage or discoloration. After the skin is gently and methodically separated from the connective tissues about the abdomen, the tailbone is cut through. Next, the skin is wriggled over the shoulders, and the wings separated from the body. The preparer lifts the skin delicately off the neck and head, minding the ears, and loosens it from the skull. The tendons that hold the eyeballs in their sockets are clipped, the eyes are carefully removed (puncturing them might produce fluid) and then the tongue. The brain is drained, and, finally, the neck is separated from the skull. The clean, dry skin is most often stuffed with cotton wool or polyester batting and wood fiber. The belly is stitched closed. The skin about the eye is pulled slightly apart, allowing some of the white stuffing to peek vacantly out. Wings are sewn in place close to the body, and the feet are crossed and tied with thread. A detailed label is attached to one of the legs above the foot, and the museum specimen is complete.

The truly excellent preparation of a study skin is a difficult skill, requiring patience, delicacy, experience, and neat stitching. In spite of my animal-rightist tendencies, I prepared three specimens while I was in graduate school (a House Sparrow and two tiny Japansese quails) before retiring from skin preparation for all time. If I may say so, one of the quails turned out quite well, though the other two birds were lumpy in the wrong places and later began to smell.

The vast majority of Darwin's ornithological specimens from the *Beagle* journey are missing. John Gould made regular personal use of the prepared skins, giving them as "special gifts" to those subscribing to his expensive bird books. By Gould's hand, Darwin's birds passed to institutions around the globe, with no good record kept regarding what went where. To complicate matters for future historians, Darwin requested that his remaining specimens be transferred from the Zoological Society to the British Museum, just as the latter institution was in the midst of moving its collections to a new

London building. In the chaos, little attention was given to the registration of Darwin's collection. The haphazard removal and replacement of Darwin's original labels (which, though simple and often riddled with error, were at least in the famous naturalist's hand) further diminished the collection's usefulness and authenticity. Most of the few remaining specimens now reside at the British Museum of Natural History's branch at Tring, and they are in just the shape one would expect of 170-year-old, roughly handled, arsenic-treated bird skins.

While it is clear that the progression of nineteenth-century ornithology depended on the careful creation and preservation of working collections for study and comparison by competent naturalists, the fact that such collections are still maintained and pursued comes as a surprise to many. In the best natural history museum presentations, a visitor today might see tropical birds arrayed in all their colorful wonder, eggs, nests, or dioramas of local birds in their proper habitats. Modern exhibits bear little resemblance to their nineteenth-century precursors and truly do have educational value. Most museumgoers assume that these displays of fine taxidermy *are* the museum's collection, but in reality the true collection — thousands of cotton-stuffed bird bodies — lies in drawers in the bowels of the institution. There are at least nine million such bird skins in the United States, some of them well kept, some of them disorganized, some of them falling, like their friends of the previous era, into insect-nibbled bits.

These hidden collections are not just relics of a bygone time. While some academic natural history museums do just barely manage to maintain their sad, dusty little collections that have been hanging around for decades, many of the more well-endowed are collecting actively. In Seattle, for example, the University of Washington's Burke Museum maintains one of the best collections west of the Rockies, including some sixty thousand birds.

When I was helping to organize the Seattle Audubon Society's Master Birder program (yes, "Master Birder," a program providing opportunities for advanced birders to deepen and round out their avian education), I invited Dr. Sievert Rohwer, the Burke Museum's ornithological curator, to give a talk called, "Living Biology from Dead Birds," about the use of museum specimens in modern science. Dr. Rohwer himself has undertaken some of the most recent, and most elegant, studies involving ornithological specimens. He and his students have collected Townsend's Warblers, Hermit Warblers, and their hybrids in the Pacific Northwest. Their research has uncovered a trove of hitherto unknown aspects of the hybridization process, the creation of hybrid zones, and the "personality" traits of particular species insofar as they affect hybridization. During his talk, Dr. Rohwer showed slides depicting the range of hybrid characteristics that emerge when these species mingle. There is an almost linear progression from a pure Townsend's to a pure Hermit, with their hybrids spread progressively between them.

Obtaining such specimens required a bit of ingenuity. The researchers set up tiny stuffed specimens of Townsend's or Hermit warblers on stands in appropriate habitat locales during the spring (when the birds were singing to establish territories and behaving with heightened aggression, fending off uninvited birds from their own would-be nesting places). A recording of the appropriate bird's song was played next to the little manikin, and when a warbler flew in to chase off the fake marauder, it was shot by a hidden ornithologist. Rohwer held up prepared skins as he made various points, and he spread about twenty of them, all arranged as they had been on the slides, on a table for viewing. Master Birders filed by, while Rohwer stood guard, making sure the tea and cookies handed out at break time were not dribbled over his specimens.

Rohwer's research made fascinating science, and it was one of the most captivating programs on the whole Master Birder syllabus. But several of the students had mixed feelings, and one woman walked out. Quietly, with no ruckus, but out nonetheless. After all, the people present were in the program because they loved birds. Surely there was plenty of excellent and necessary ornithological study to be accomplished without luring innocent warblers to their deaths.

Many people, and not just those in the uninformed general public but also scientists of various sorts, conservation biologists, even some ornithologists, believe that the continued collecting of avian specimens is unnecessary, unethical, and a throwback to nineteenth-century methods that we should have long outgrown. Much of the information that collectors obtain from specimens might be gathered from a live bird in the hand, one that has been mist-netted (caught in a large, lightly strung net) and can then be released, following the method of bird banders. Data such as location, elevation, weather, and habitat of a captured bird clearly do not require a bird's demise. Length, weight, skull ossification, degree of molt, color and measurements of bare parts, reproductive readiness, and the presence of ectoparasites all might be recorded by a skilled observer of a live bird. Even DNA, blood, and tissue samples can be gathered in the field without undue trauma to even the smallest birds. Surely the nine million birds already lining museum drawers across the United States are quite enough. Can't we turn our attention now, in a more enlightened millennium, to the living?

But collectors counter. Not all birds can be mist-netted or otherwise captured. Many are too large, ungainly, clever, or situated beyond the reach of even the most intrepid and physically adept researchers. It can take all day to mist-net just two or three of the target species, while twenty can be "taken" on a good day of collecting; given the drastic limitations in both time and funding that haunt university studies, live capture is an unattainable ideal. In a good collection of specimens, they argue, examples of species in an entire family — an entire order, even — might be laid out together, compared as individuals, and then viewed within a continuous lineage. The creative leaps that encounters with such specimen series might spark in a scientist are impossible to quantify and could inspire some of our best future research and understanding. And their *pièce de résistance* is difficult to argue with. We had no idea until very recently that we would be able to learn about nineteenth-century avian populations by sampling the DNA of older specimens. Who knows what the next millennium will require of today's stuffed birds? The further reaches of ornithological science are unimaginable, collectors say, and scientists have a selfless obligation to provide material for future researchers.

The issue of capturing versus collecting is complicated by misunderstandings that flow both ways. Many who are against collecting in the name of protecting bird lives fail to realize the extent to which museum collections can contribute to the conservation of species and ecosystems. Such collections help us determine species and subspecies, which in turn allows us to catalogue biodiversity and create the geographical range maps for species that underly conservation decisions. They allow us to monitor changes in populations and evolutionary processes that might alert us to conservation priorities. They aid in the education of young people who will become the future voices for conservation. The illustrations that are the heart of the field guides we all carry could never have been rendered without the artists' constant reference to museum collections.

But ornithological curators tend to oversimplify the motives of those who question the ethics of continued collecting. When I asked Dr. Rohwer what he thought about the feelings displayed by some of the Master Birders, he told me that "Audubon people," and people in general, for that matter, tend to suffer from a "Fragile Nature" syndrome. Modern urbanization has distanced most of us from the rampant mortality that characterizes natural systems. We are oversensitive to the deaths of individual creatures in this vast system, where nature's fecundity is matched by its harshness. Alongside the natural deaths in avian populations wrought by inclement weather, parasites, predation, genetic weakness, and bad luck, as well as the human-caused deaths from cars, pollution, habitat destruction, and domestic cats, the number of birds affected by museum collectors is statistically insignificant.

I think Rohwer is absolutely right. Nature remains romanticized, and therefore misunderstood, largely because we live such coddled lives at such a remove. But many thoughtful and scientifically grounded people have their eyes wide open in the face of nature's tooth and claw yet remain sensitive to the deaths of individual birds. The collector's perennial argument, that conservation is accomplished at the ecosystemic level rather than at the individual, is mostly true. The conclusion drawn from it — that ethical concern over the killing of birds for collections is therefore misguided — is not. Humans are complex beings, perfectly capable of dimensions of concern. We are able to comprehend the overarching function and structure of the ecosystem while remaining interested in the well-being of creatures, our fellow beings. We are right to contemplate, as deeply and as well as we can, the ways in which humans interact with nonhuman lives, particularly when the killing of animals is involved. This is not to say that birds should not be collected, but only that those who argue against it are not necessarily scientifically ignorant.

I should confess at this point that I have suffered a long-standing bias against collecting, and even educational taxidermy, feeling that students would do better studying good photos of live birds, and of course birds in the field. I remember even as a child squinting at the mounted birds in museum displays, finding them campy at best. In my opinion, even the highest-quality taxidermy somehow makes a bird look more dead than one lying in the road — wired into some not-quite-right position, feathers ruffled, false eye shining bleakly. When I began working at Seattle Audubon, we had some very bad stuffed specimens lining the shelves and gracing the tops of file cabinets. A goldeneye, a puffin, something that used to be some kind of shorebird, and everyone's favorite — a tall Great Blue Heron, actually named Harry. They were all dusty, and old, and completely horrible. But for some reason, most of my officemates were quite attached to them. I was new to the office and had not bonded with the stuffed beasts as the others seemed to have done. I enlisted another relative newcomer into the cause — my neighbor Helen in the next office — but whenever we suggested that the unfortunate stuffed birds might do better in the basement, the others would shrug noncommittally and say, "Well, I don't know. They've always been here." One day Helen sneaked conspiratorially into my office and whispered, "Good news! A new volunteer is sensitive to dust mites and thinks the taxidermy is aggravating her allergies!" Helen and I graciously volunteered to wrap the things in plastic and relocate them, unceremoniously, downstairs.

But out of sight is not necessarily out of mind. Teachers kept calling to request Harry's presence in their classrooms. "We've used him in the past as an inspiration for art projects," the teachers would plead. Art? The children would draw a bird, with Harry as their honored model. "We don't really loan the bird out," I'd say. "He's frail and dusty. Children with allergies might react badly." But the teachers would persist — they had borrowed him before. Clearly I was new and didn't know better. They would come to pick up Harry tomorrow. "But wait," I'd try again. "He's not really that inspiring. He's very dead. Surely a living bird would make a better muse." Perhaps the zoo? Perhaps a bird-watching trip? I would even lead it — anything to keep poor impressionable young minds from conflating the dreary stiffness of dead birds with the astonishing wonder of live ones.

But the dead birds kept insinuating themselves. Dedicated Audubon members continued to bring fresh specimens to the office — a Cedar Waxwing that hit the window, a Wilson's Warbler brought in by the cat, a Dunlin found in a cornfield while bird-watching, a Varied Thrush mysteriously dead in the birdbath. As the education person, I held the governmental salvage permit that allowed us to keep such birds. They lay within my purview. I wrapped them in plastic and stuffed them into the freezer section of our small office refrigerator, thinking that one day I would find a graduate student at the University of Washington to prepare them for me. Maybe I would have this imaginary student "part them out," as they say, into outstretched wings, bills, and feet that teachers might use if they persisted in their morbid tendencies.

But then my officemates began to complain. The freezer

was overflowing; there was no room for frozen yogurt or Frap-
pacino pops. So one day I bundled all the birds up and took
them home to my larger freezer. We had added a very nice
crossbill, a Great Horned Owl, and, sadly, a beautiful adult
Sharp-shinned Hawk that had been killed by a car.

One night, about to serve up a nice ice-cream dessert to
some dinner guests, I decided that as long as I was moving the
owl to get to the Häagen-Dazs anyway, I might as well bring
her out and show her around. Owl feathers are covered with a
soft downy fuzz, an adaptation to cut down on the noise of
wind resistance, giving the owls silent night flight. Because of
this, even frozen owls are incredibly soft, as soft as chinchillas.
Still, I was not prepared for the reaction of my guests, who
went absolutely wild. They suddenly lost their normal voices
and began to whisper, they rubbed the owl against their
cheeks, they exclaimed over her feathery eyelids, they brushed
her smooth breast with the backs of their hands, they carefully pressed their fingertips against her sharp talons and her
hooked beak. I leaned against the doorframe with my mouth
open, watching.

The very next day I called the Burke Museum, enlisted a
willing volunteer, and eventually ended up with an enormous
box full of wings, feet, and skulls (one of a screech owl, showing the asymmetry of owlish ears; another of a flicker, showing how the tendons that attach a woodpecker's tongue reach
all the way around the back of its skull and attach to the top
base of its bill) and many boxes full of study skins. No creepy
mounted display taxidermy, thank you, so my box of fresh
new specimens was quite stunning. When I opened it for
schoolchildren, they went absolutely crazy, even crazier than
my owl-loving adult friends. Every little bird piece was a new
wonder. Even individual feathers seemed to astonish them.

For the children, the experience of examining specimens did not compare with birding in the field or — even more amazing for them — banding live birds. When banding, a child can feel the beating heart and see the bright eye, so complete, so small, and hold an entire life in the palm of her hand. And after all the measurements have been taken and all the information recorded, when the tiny metal band is in place and the bird is set lightly in the palm of a child's hand to fly free, the child can imagine that the bird waits that long moment in her hand, not because it is scared witless and has forgotten in this one second what to do, but because it somehow *wants* to linger there with her. It is a harmless thought, and irreplaceable. So no, no specimen can offer such an experience, but it can, I finally learned (more slowly than everyone else), enhance it. Specimens do, in the best circumstances, come to a new kind of life.

When I interviewed Sievert Rohwer, he greeted me warmly, with twinkly eyes. I immediately felt bad for quipping to my husband as I walked out the door that morning that I was "off to talk with the warbler killer." Very generously, Dr. Rohwer spent the entire morning in his office with me, talking about birds and collecting and science — not just about Pacific Northwest warblers but about Russian wagtails, extinct heath hens, and African weaver finches.

The Burke Museum's collections are housed in a large room lined with enormous white metal cabinets that look like giant refrigerators. The doors open to reveal rows and rows of drawers filled with birds. They are all lined up tidily, their little dried bills pointing in the same direction, their necks and bellies exposed, their feet crossed and tied neatly together with black thread. The whole effect is quite morgue-ish. Some of the labels attached to the specimens' legs are computer-generated and new looking. Some were scripted in fountain pen a hundred years ago, or more.

A good specimen label will proclaim the bird's scientific name; the date and place it was collected; its age and sex, if known; and perhaps other tiny facts regarding its life and provenance that are pertinent and succinct enough to fit on such a slender swatch of paper. For all of Darwin's stringent record keeping on the *Beagle,* he never managed to make proper labels like these for the zoological specimens he gathered. Rather, Darwin employed the labeling method practiced within the discipline he knew best — that of geologists and paleontologists. Because it is difficult to affix a label to a stone or fossil, such specimens are discreetly numbered with India ink, and the number is cross-referenced in a notebook, where detailed information is recorded. But what is sensible for rocks may not be so for birds, and it seems odd that Darwin never caught on. Most field ornithologists, including Darwin's contemporaries such as Alfred Russell Wallace, attached detailed labels with string to the bird's feet, appending only lengthy observations in a notebook.

Dr. Rohwer and I explored the tropical tanagers, the spinetails from Darwin's notes, the Northwest thrushes, and the orioles. I held a female and a male Ivory-billed Woodpecker, recently extinct. Rohwer looked at me a little uneasily when we came to the warbler cabinet. "I don't want to shock you," he told me kindly before opening the drawer to expose a sea of black, yellow, and white bodies, the drawers above and below brimming with more of the same. He showed me all the characteristics that it is impossible to distinguish in the field, without a specimen, and showed me a line of warblers that, to any birder, would look like a Townsend's, while their DNA reveals that they are Townsend's-Hermit hybrids. My own response confused me, for I was enthralled by the science and appreciative of Rohwer's own enthusiasm for his work, but I was also carried away, reluctantly, by the beauty of these birds. Not the life that I imagined for them, their wild lives, but their presence here, just as they are, their scientific lives.

I asked Rohwer how many warblers were collected for his studies. He hadn't counted but could guess. "About a thousand," he reckoned. I nodded and repeated the number to my sister, a physics teacher, that evening. "A thousand?" She was surprised. "Is it worth it?" she asked. "I don't know," I told her. And I still don't.

*D*arwin's specimen catalogue divides his collection into two large basic categories: *Specimens in Spirits of Wine,* and *Specimens Not in Spirits.* There were 3,921 specimens in all, most of them prepared during the *Beagle* journey with the unflagging and indispensable help of Syms Covington. The majority of the several hundred avian specimens lay in the "Not in Spirits" category — the usual cotton-filled, arsenic-dipped, dead-looking, slightly too-fat-or-too-thin specimens that have lined museum drawers for the past couple of centuries. "Specimens in Spirits of Wine" were typically creatures that looked tolerable when wet — frogs, fish, lizards, insects, corals, snails, worms. Jar number 213 was a "large bottle, full of spiders." Number 134 was a fungus, found "growing on a wet plank in a darkish outhouse," and number 375 was an armadillo. But there were also a few avian specimens on this list. Most of these were bits of birds, dissected out and preserved for later study. The first bird-related specimen, though not actually part of a bird, was number 400, "an intestinal worm taken out of the stomach of an Ostrich." Number 508 was the "trachea of a Goose," and number 620 was the tongue of a woodpecker.

While most complete birds were made into skins, a couple were preserved by Darwin in "spirits," or ethanol. The first of these was in jar number 418, a "Tringa" or long-legged shorebird, "shot out of large flock," early in the voyage at Bahia Blanca. The next was number 630, Darwin's loosely-plumed reedhaunter — the same as dry specimen 1228, he noted, "only with a tail." Excepting his experience with beetles, Darwin was new at collecting. He experimented. While he was surely pleased with his frogs and corals in spirits, the birds he pressed into jars must have looked disappointingly wet.

I wonder, in fact, whether all of Darwin's specimens looked a bit more dead than he had hoped. Here he was, re-turned from his journey, attempting to fashion his experiences into scientific respectability. But was this what it meant? Rows of bird skins with dusty cotton coming out of their eye sock-ets? What had these to do with the little Turca bird that had run before him, glancing backwards and lifting its tail? Where, between the conveniently still specimen on the table and the species that Gould described in a list of measurements, did the living bird have its say? Perhaps Darwin saw that the spec-imens could not convey what he had learned to be true — that an organism's preciousness lay not only in its description but also in its life. Darwin wondered how he could ever trans-late his private understanding meaningfully — being true to both the requirements of the science of his time and his own experience. The dissonance within him surely deepened.

But Darwin pressed forward. He worked busily to rewrite his diary as the *Journal of Researches*, which would be published alongside FitzRoy's own labored account of the voyage. In the *Journal*, Darwin followed his diary carefully, but from the vantage, always, of the finished journey. His mind wandered forward and back in time as he conflated certain passages and tampered with chronology in order to create a readable book. It was significant, this setting on paper, this movement of his voyage from a disparate jumble of experiences into a contained, readable, somehow finished work. For Darwin, the *Journal* served several functions: a public presentation, a scientific entrée, and a psychological ordering.

Darwin received a governmental grant of one thousand pounds toward the organization, description, and illustration of his collections in a multivolume set titled the *Zoology of the Beagle*, with each volume covering a particular zoological category — reptiles, fish, mammals, and, in volume three, birds. Each species was described and named and all were arranged according to the current taxonomic standard — or at least that favored by the naturalist in charge. Darwin's notes regarding range and habits were included, adding a great deal of charm to what would otherwise have been a very dry compendium. The finished books were lovely, with green leather covers and colored lithographs.

Now the neatly numbered and labeled specimens were arranged happily in proper drawers and cabinets, and they were impressive both in their sheer number and their scientific weight. Darwin had a stack of books, too, as well as the *Journal* and the *Zoology*, and they were selling well. There must have been a great deal of satisfaction in seeing the sum of his work and experiences presented so impressively between covers. But there must also have remained, for Darwin, the realization that the journey's center could never be contained in this way. Looking quietly upon his avian specimens, Darwin saw their lives in them, the interaction of their lives with his own, and with life beyond. He had no way to translate this sense into proper language within the confines of a text like the *Zoology*, but he could allow its validity in the shape of his thought. He carried a notebook in his pocket, a secret one, and it grew from all of these things — these skins and books and jars — and also from that dimension of Darwin's mind and experience that was entirely his own. There is no doubt that the ordering of his specimens in drawers and in print created both a psychological closure and a deep sense of accomplishment for Darwin. But the things they didn't, or couldn't, contain were the very things that compelled Darwin forward in his own life and work into fanciful experiments, wild sidetracks, compulsive single-mindedness, barnacles, ulcers, pigeons, orchids, earthworms, and natural selection.

CHAPTER ELEVEN

A Naturalist's Fancy

I have been led to study domestic pigeons with particular care. . . . The details will often be tediously minute; but no one who really wants to understand the progress of change in domestic animals, and especially no one who has kept pigeons and has marked the great difference between the breeds and the trueness with which most of them propagate their kind, will doubt that this minuteness is worth while.

— *The Variation of Animals and Plants Under Domestication*, 1868

My dear Fox," Darwin wrote to his beloved cousin in the spring of 1855, "I forget whether I ever told you what the object of my present work is, — it is to view all facts that I can master (eheu, eheu, how ignorant I find I am) in Nat. History, (as on geography, distribution, palaeontology, classification Hybridism, domestic animals &

plants &c &c &c) to see how far they favour or are opposed to the notion that wild species are mutable or immutable . . ." He was partly disingenuous here in referring to such efforts as his "present work," for certainly he had been about such tasks for nearly two decades.

By this time Darwin had managed to round out his life in other, less scientific ways. He had married his first cousin Emma Wedgwood, an intelligent and comfortable woman, musical and literate, slightly frumpy, religious but not pious (rather more Anglican than the better part of the Unitarian-leaning Wedgwood clan), and indulgent of her chosen mate's eccentric scientific tendencies. These she considered more of an elaborate hobby than high science, and in spite of Darwin's eventual fame, she would never entirely abandon this view. They were married early in 1839, and occupied rooms on Upper Gower Street in London. They dubbed their first home together "Macaw Cottage," in homage to its frightful décor — blue walls, yellow drapes, and overstuffed red velvet furniture. They reproduced — William Erasmus was born in 1839, and favorite daughter Anne Elizabeth in 1841. Darwin set immediately to work studying the habits and facial expressions of his infants, comparing them to the baby primates at the local zoo. He was a doting and affectionate, if preoccupied, father.

In 1842, Charles and Emma bought Down House in the country village of Downe, about sixteen miles southeast of London, where they would live the rest of their lives. Days after the move, Emma gave birth to a quiet, sickly girl, Mary Eleanor, who would die just three weeks later. By the time Darwin wrote to Fox in 1855, Emma had borne six more children, and would bear yet another in 1856, a weak little boy named Charles, who lived just three years. Though Darwin was heartbroken at the loss of these two babies, it was the death of his most beloved daughter, Annie, at age ten, that was the defining tragedy of his life. He was with her throughout her agonizing illness, he wept at her bedside as she died, and he never recovered from her loss. Any lingering traditional religious sensibility he might have harbored was buried with little Annie as he repeated to him-

DOWN HOUSE

self the words uttered by countless thousands of grieving parents over the millennia who find no comfort in the rote words of a clergyman. *No God would let a dear child suffer in this way.*

During all of this time, Darwin kept his notebooks on the transmutation of species. His reading of Malthus's population studies helped him to solidify his own theory of natural selection as the mechanism of evolutionary change; in 1842 he wrote a thirty-page "sketch" outlining his basic ideas, which was expanded into a 230-page "abstract" in 1844. He had a fair copy of this last made and wrote a sealed note to Emma, giving detailed instructions for its posthumous publication should he be thrown mortally from his horse.

In this same year, Darwin was blindsided by the publication of a book explaining earthly life through evolutionary change, titled *Vestiges of the Natural History of Creation.* Written by the Edinburgh journalist and publisher Richard Chambers, *Vestiges* was published anonymously, popularly read, and met with simultaneous excitement and revulsion. The invisible author was slammed as a "practical Atheist," and, worse yet, Darwin's own scientific circle despised the work as ungrounded rubbish that never got around to improving upon Lamarck. Not only was Darwin crabby about the fact that he was not the sole English gentleman engaged in the earnest consideration of evolutionary processes (as he had seriously, perhaps foolishly, considered himself to be), but he was taken aback by the vehemence of the criticism aimed at the book. *Vestiges* was not as astute scientifically as Darwin's work on the subject, and it did not present a cohesive mechanism of evolutionary change as Darwin planned to do, suggesting instead a vague sort of physical "tendency" in organisms to evolve over time. Still, the book did gather in one place the same sorts of evidence from the same branches of natural history that Darwin had been working on in secret for the past decade. Chambers drew on geology, zoology, embryology, and animal behavior to buttress his conclusions, which were — at least in their broad strokes — very similar to Darwin's. Darwin felt nauseous as he imagined all the vitriol engendered by *Vestiges* being turned upon his own work, which would not be anonymous. To make matters worse, he felt a pang of jealousy, difficult to admit even to himself, over the plain, outspoken courage of this author — veiled beneath anonymity though it was — a courage Darwin found wretchedly lacking in himself.

Darwin listened awkwardly as his colleagues, Lyell and Hooker among them, vilified "Mr. Vestiges." Clearly the man was not a proper natural historian. His geology was superficial, his zoology was murky, his classification was dubious. Darwin's nervous disorders worsened as the plague of self-doubt deepened within him. He felt secure in his grasp of geological and zoological principles, but he relied entirely on others for expertise in classification. Was he no better than this *Vestiges* interloper? He would make sure that he was. The publication of *Vestiges* was one of the elements that pressed Darwin into his most obsessive undertaking to date.

When he began it, Darwin did not intend to write a twelve-hundred-page monograph on the classification of the world's barnacles. He meant only to tackle one or two of the more difficult groups. But the project grew out of hand, as such things can do, and Darwin worked at the monograph for eight years, to the near exclusion of all other research. He called in specimens from institutions, naturalists, and friends. He wrote to Syms Covington in Australia, with whom he had not communicated in years, and after a polite summation of family interests and inquiries announced his true purpose. He did not know whether Covington lived near the sea, but if he did, Darwin "should be so very glad" if Covington would collect for him any barnacles "that adhere (small and large) to the coast rocks or to shells or to corals thrown up by gales, and send them to me without cleaning out the animals, and taking care of the bases. You will remember that barnacles are conical little shells, with a sort of four-valved lid on the top." In classic Darwin fashion, he concluded, "I should be very glad of any specimens, but do not give yourself much trouble about them."

Walling himself in his room, coming out only to tousle the hair of his children, sneak snuff, and await the post, Darwin became consumed with barnacles, as did his entire household, until the whole situation seemed almost normal. "Where does your father do his barnacles?" one of Darwin's children asked innocently upon visiting a friend's house. Poor thing.

The resulting *Monograph on the Sub-Class Cirripedia* was beautiful, illustrated in the fine line drawings common in scientific publications of the time, and wondrous in detail. More than

150 years after publication, it remains useful as a scientific reference. Still, I imagine it is altogether possible that ever since Stephen Jay Gould passed away, not a single living being can honestly claim to have read the entire thing. By the time he was finished, Darwin's health was worse than ever, he was addicted to snuff, and his eyes seemed to bulge. But at least no one could say that he lacked experience in taxonomy and classification. Finally, he could stretch his tentacles into the many facets of natural history research that had been calling to him, as if through a fog, during the barnacle years.

Darwin threw himself into the question of dispersal, the movement of organisms from one place to another, particularly organisms that could not of their own volition swim or fly. Organisms like plants. Certainly Darwin wanted to counter the idea of some hapless God having to busy himself with the landscaping of the entire earth; but he also hoped to do away with the then-prominent scientific notion that historic land bridges, now submerged, once connected islands to major land masses. These bridges, it was supposed, allowed the dispersal of plants as their seeds were blown along, and even very slow animals like slugs and spiders, which, if they kept at it, could eventually emigrate across the hypothetical land bridges. Darwin had been tracking and mapping the presence of various plant species across the face of the earth, and he had come to believe that land bridges were theoretically superfluous. Many organisms, he thought, possessed a powerful tendency to vastly disperse themselves. He began to obsessively conduct a series of experiments designed to determine the means by which a nonswimming, nonflying organism might contrive, within its seemingly humble means, to get itself moved across land or, a more trenchant inquiry, across water.

Darwin set up aquariums full of saltwater and floated a variety of seeds in them. Some sank, but many floated and, when retrieved and planted, germinated. His cabbage seeds germinated even after twenty-one days in the brackish water! Darwin was exultant. That was plenty of time for a seed swirled by ocean currents to travel a fair distance. He crowed the good news to his dubious colleagues. For the most part, Darwin's plant seeds did better in his smelly aquariums than the reptile and mollusk eggs he floated, but several of those survived as well.

Still, Darwin realized that dispersal could not be explained by sea travel alone, and soon his studies drew him back into the sphere of bird life, inviting him to look again at the place of birds within the movement of life. He came to suspect that, besides such rafting, dispersal via birds was the most common method for plants, and even some small animals, to colonize distant lands. There were various ways that birds might carry tiny unknown passengers, and Darwin began to explore them all.

He fed seeds to the seabirds at the zoo, then came back to collect their guano, plant it, and observe which seeds germinated. He sent his devoted butler Parslow out on rainy days to shoot partridges so he could peel the mud off their feet and see how many seeds it contained. He instructed another fortunate servant to wander the poultry yard with a keeper, wash the feet of all the partridges, and save the dirty water, which he would sift for botanical evidence. Not satisfied with his own grounds, and unable to travel himself due to ill health, Darwin shamelessly pestered his Cambridge contemporary Thomas Campbell Eyton:

*I want to know whether on a wet muddy day, birds' feet are dirty.
But I want especially to know whether herons or any waders or
water-birds when suddenly sprung have very dirty feet or beaks:
Do you know when owl or Hark eats little birds, how soon it
throws up pellet? Can it throw up pellet whilst on wing? How I
should like to get a collection of pellets and see whether they
contained any seed capable of germination. Could your
gamekeepers find a roosting place, and collect a lot for me?*

In his dispersal studies, Darwin perfected his lifelong habit of
asking others the most extraordinary of favors while making it
seem as if it were he himself who was favoring his correspondent with his attention. As always, he signed off charmingly,
"Do pray help me with your advice, and forgive this trouble.
Yours very truly."

It was common for Darwin to involve the children in his
inquiries. One time, he cut the legs off a dead duck and let the
children wiggle them around in his aquariums, as if they were
attached to a swimming bird, to see whether the freshwater
snails he was raising would cling to the feet. The snails obliged
nicely, nestling between the webs in numbers. Next, the children waved the feet through the air, imitating flight, then rejoiced with their oddity of a father as the little mollusks held
tight and didn't begin to dry out and die for nearly twenty-
four hours — time enough for a bird to fly seven hundred
miles, Darwin reckoned, smiling to himself. Blown off course
by a storm, one bird could populate an entire island with a
new species of snail. Darwin's conclusions about botanical
distribution, fabulous, homespun, and fresh in his time, are
now basic ecological tenets.

Alongside dispersal, a vast array of subjects, all of them
turning somehow around the twin subjects of evolution and

natural selection, occupied Darwin's mind and studies. But, "believing that it is always best to study some special group," he wrote early in the *Origin*, "I have, after deliberation, taken up domestic pigeons." For Darwin, who possessed a beautiful flair for abandon, "study" as it appears here was a code word for "immerse himself obsessively." Over and over, throughout the whole of his life, he would throw himself with an intrepid fearlessness into a subject that drew him. His lack of a sense of moderation in such matters is a condition for which millennial humans might be medicated. But for Darwin, the capacity for full immersion was one of his defining traits as a thinker, and as a naturalist, one that shaped his philosophy and worldview.

I guess there is a kind of inevitability in his choice to study pigeons. Darwin had, for some time, sensed the relationship between the role of domestic breeders, who selected for certain traits in their animals, and the place of natural selection in the wild. He studied cattle, pigs, dogs, and all manner of poultry, but he found in the pigeon the perfect subject. There was no other domestic creature that evidenced so much variety, and all of this variety was accomplished, Darwin strongly believed, by pigeons as descendents of a single species — *Columba livia*, the common Rock Dove of city parks and McDonald's parking lots.

The point was an important one for Darwin. The notion that a particular lineage could achieve myriad forms was one of the thrusts of the *Origin's* long argument. If all of the beautiful breeds of domestic pigeons, with their incredible variety of plumage, bill length, tail length, and even number of vertebrae, were examples of *Columba livia* (just as both Chihuahuas and Newfoundlands are breeds of the same species, *Canis domesticus*), then he had in hand a perfect example of the twining of both variety *and* identity within a lineage. In the *Origin*, and in his later *Variation of Animals and Plants Under Domestication*, where he presented his domestic studies in two volumes of marvelous detail, Darwin took care to outline his argument that all fancy pigeons were, at heart, not-so-fancy pigeons.

The great majority of ornithologists believed the same thing — that all fancy breeds were merely extravagant representations of their parent species, the Rock Dove. Pigeon fanciers, obsessed with the collecting and breeding of pigeon varieties, strongly disagreed, convinced that the objects of their infatuation were descended from several interesting, unknown, or intriguingly extinct wild ancestors. This is just one example of the various intellectual and social gulfs that separated the ornithologists and the fanciers, who generally had little use for each other. The members of the pigeon fancy found the ornithologists to be high-minded, dismissive, and condescending. The ornithologists simply had no need for the fancy — once they'd reduced all the pigeon breeds to Rock Doves, there was little left, after all, of ornithological interest.

Darwin wasn't the least interested in the fanciers' indefensible views of pigeon origin and descent. They were free to believe what they would, misguided but harmless. He was, however, interested in them for their eye. Here in the fancy, Darwin found his opportunity to actually watch the movement of selection unfold before him, but he recognized that he couldn't *see* pigeons well enough to fully understand the working of the breeder as a selecting agent. What did the experienced breeder see when he looked at two birds of the same breed, birds that looked for all the world to Darwin to be exactly the same, and pronounced one of them to be superior to the other? What subtle differences in length of bill, plumage color, feather luster, posture, bone structure, or nostril size did the fancier discern? Darwin could pick out the slightest bulge in a barnacle's ridge that would place it in a unique genus, but these dancing, feathered beings were another sort of beast altogether. They were beautiful. There were breeds, such as the Almond Tumbler, that certainly caught Darwin's fancy. But the delicate discernment of the best breeders eluded him. Darwin craved this eye for himself, and he attacked the subject from all angles.

In London, the pigeon fancy was, and still is, a thriving avocation. Historically, the fancy had grown up around taverns in the city's seedier underbelly, but by Darwin's time, the hobby had gained a kind of respectability. Working-class aspiring gentlemen made up a portion of the fancy, a scattering of aristocrats kept birds, and even the young Queen Victoria had a beautiful dovecote built and filled with her favorite breeds. For the most part, though, a social chasm separated the gentlemen natural historians from the more plebian pigeon hobbyists, and most of Darwin's scientific colleagues would not have been caught dead at meetings of the pigeon clubs that their eccentric friend now attended regularly. Darwin joined two clubs devoted to the fancy, including the most pretentious of the London organizations, the Philoperisteron Society, which required introduction by a member, near-unanimous election, and prohibitive dues. But even here he was rather out of his element and consciously distanced himself from the full social dimension of the clubs. He enjoyed playing up the irony of his association among friends. "I have now a grand collection of living & dead Pigeons," he wrote to the American zoologist J. D. Dana, "& I am hand & glove with all sorts of Fanciers, Spital-field weavers & all sorts of odd specimens of the Human species, who fancy Pigeons."

The pigeon is one of the best-documented birds in all of

history. The first known pigeon clubs date to the 1750s, but a full four hundred years of art and literature portrayed fancy breeds and the wild birds from which they might have descended. Darwin pored over it all, carefully tracing the historical changes that led to modern varieties. "I am making some progress," he wrote to dear Fox about his ancient pigeon studies, and "was yesterday in the British Museum getting old Chinese Encyclopedias translated."

Darwin attended pigeon shows where many of the best breeders displayed the treasured gems of their flocks. The annual show at Birmingham was the largest, the "Olympic Game of the Poultry World," as the London periodical *Cottage Gardener* referred to the exhibit in its weekly poultry column. Birmingham helped highten awareness of the fancy in the mind of the general public, and families with picnic baskets came to enjoy the views and the music, letting their children run loose between rows and rows of exotic pigeons. Darwin's own Philoperisteron Society organized two well-attended shows, one in winter for adult birds, and one in the spring for young. Darwin myopically scrutinized the birds upon which the judges conferred their coveted prizes.

Eventually he brought his new obsession home with him, hiring a local carpenter to build a state-of-the-art dovecote to his specifications at the back of his gardens at Down. He visited John Bailey of Fleet Street, among the premier judges and poultry dealers in greater London, and started his flock. Like so many other fanciers, Darwin fell in love with the *au courant* breed, the Almond Tumbler, not the fanciest pigeon but lovely in the delicate proportions of its face and bill. Darwin delighted as his "little flock" grew and changed, and was very nearly rapturous as he chose new chicks to carry home on his lap in small wooden boxes. He was given rare pigeons by fanciers who were only too thrilled to rub elbows with the famous Mr. Darwin. Eventually he had nearly ninety birds and was forced to expand the coop. "I am getting on splendidly with my pigeons," he wrote to his son William, or "Dear old Willy," as Darwin called him, now away at school. "I visited a jolly old Brewer, who keeps 300 or 400 most beautiful

ENGLISH POUTER

pigeons & he gave me a pair of pale brown, quite small German Pouters: I am building a new house for my tumblers, so as to fly them in the summer."

Darwin enlisted the help of his children in caring for the birds, and thirteen-year-old Henrietta, particularly, showed interest and aptitude as a poultry enthusiast. But when her cat was found to be catching a disproportionate number of pigeon offspring, it was put to death without, it was thought, her knowing. (As so often happens, the intuitive wisdom of girls was underestimated; Henrietta knew exactly what had happened to her cat.)

Darwin's requests for pigeonly assistance from all quarters of the globe were, excepting his requests for muddy bird's feet, unrivaled in their charming shamelessness. He wrote to the prominent fancier William Bernhard Tegetmeir, who would become his primary poultry mentor, "I have been thinking over your offer of helping me to the bodies of some of the good birds of Poultry. — Really considering how complete a stranger I am to you, I think it one of the most good-natured offers ever made to me." Darwin insisted on paying for all "carriage, porterage, booking, baskets, &c" involved in the posting of the pigeon bodies and, in gratitude, offered

Tegetmeir a copy of his by now famous *Journal,* which, Darwin wrote simply, "has been liked by some naturalists." He hoped to make skeletons of various breeds for comparison, but he hated to buy birds just to kill and boil, so if his comrades had dead bodies to spare, he was always indebted.

The skeleton business was not as straightforward as he'd hoped, however. For one thing, the smell of putrefying pigeon flesh boiling down on the kitchen stove was nearly too much to bear. Darwin's delicate stomach wretched awfully, and though Emma bore the whole thing bravely, she had, for once, to disapprove of her husband's methods. He wrote again to his naturalist colleague in Shropshire, Thomas Campbell Eyton, since he "had such great experience in making skeletons." Darwin confessed his difficulties, wondering whether Eyton might be so kind as to offer some assistance:

> *Now I was told that if I hung the body of a bird or small quadruped up in the air & allowed the flesh to decay off, & the whole to get dry, that I could boil the mummy in water with caustic soda, so get it nearly clean, but not white, and pray tell me how do you get the bones moderately clean, when you take the skeleton out, with some small fragments of putrid flesh still adhering. It really is most dreadful work. — Lastly do you pluck your Birds?*

Mercifully for all involved, Darwin gave up "skeletonising" and decided to send all incoming pigeon carcasses out for proper preparation.

He persisted in making as complete a collection as possible of the skins of fancy breeds from all quarters of the earth for comparison. In this endeavor Tegetmeir was again invaluable. "If you should happen to stumble on a bird *in good plumage*, I wish you would have its neck broken, instead of cut, & then I shall understand that you think it worth skinning, instead of skeletonising." Darwin promised Tegetmeir that if he succeeded in making a good collection of skins and skeletons of domestic birds, he would, when finished with them himself, donate the whole to the British Museum.

Darwin was indebted to several prominent members of the fancy, but Tegetmeir is the only one we really know anything about. A kind of a renegade scholar, he was unusual among fanciers in that he possessed a solid background in natural history and was a fellow of both the British Ornithologists' Union and, like Darwin, the Zoological Society. He wrote several scientific papers, including one on the honeybee, from which Darwin drew heavily in his own discussion of honeybee cell formation in the *Origin*. Unlike Darwin, Tegetmeir was not independently wealthy and had to work in order to pursue his naturalist and pigeonly pursuits. He was the poultry editor for the London periodical *Field* and was involved in various journalistic projects that had to do with the domestic breeding of birds. Tegetmeir was well aware of Darwin's renown, and he was all too happy to be of any possible assistance to the distinguished naturalist. He offered Darwin an insider's connection to the fancy, and, as Darwin's correspondence reveals, a shocking amount of research assistance as well.

But Darwin, clearly more than Tegetmeir, was conscious of the social divide between them. Not only did Darwin pay all expenses involved in any of Tegetmeir's packing and shipping of bird bodies, skins, and skeletons, but he also paid him for reviewing an unpublished paper, which made him, in Darwin's mind at least, more of a paid assistant than a colleague of equal standing. Darwin wouldn't have dreamed of paying Lyell, Hooker, or Henslow for a similar task. It would have been considered rude. The situation came to a bit of a head when Tegetmeir wrote a piece for the *Cottage Gardener* and referred to some work he was doing *with* Mr. Darwin. Surely Darwin was to blame for any such misunderstanding, as he always made every pretense of treating Tegetmeir as an equal when they were together and behaved as though he were the center of Darwin's universe whenever Darwin wanted something from him. But Darwin was uncommonly quick in setting the record straight regarding any sort of chummy research relationship between himself and William Tegetmeir. "Mr. Tegetmeir is a very kind & clever little man," Darwin wrote to Fox meanly, "but he was not authorized to use my name in any way, & we cannot be said to be working at all together." Tegetmeir remained his most valuable connection to the fancy.

In spite of Darwin's connections with insiders, he himself remained an outsider in the fancy circle. He was, after all, never interested in perfecting and showing his own birds. He was just as interested in the relationship between the fanciers and their birds as he was in the birds themselves. He came to the fancy as a naturalist, and he observed the relationship between the fanciers and the birds from this absorbed perspective. When he began his in-depth pigeon studies, Darwin's theory of natural selection was already well formed. What the fancy gave him was an analogy to work from as he attempted to convince his naturalist colleagues of the theory's truth. The selecting hand of the fancier, Darwin argued, mimicked the working of selection in nature. And the resulting variety of pigeons, all standing neatly within *Columba livia*'s lineage, beau-

tifully exemplified the ability
of a species to change over
time, under selection's influence.

Darwin tried to distin-
guish between what he called
"methodical" selection and
"unconscious" selection. A

ENGLISH FANTAIL

fancier worked methodically, selecting certain traits such as
color, bill length, and tail size that he hoped to emphasize in
the offspring of the matches he arranged. But the resulting
changes in the lineage over generations would be influenced by
a dynamic relationship between a whole variety of morphological and internal characteristics of the birds involved,
changes that could not be foreseen by a fancier but that could
be drastic in their cumulative effect. The difference between
methodical and unconscious selection was not Darwin's clear-
est point but he needed to make it. By making the analogy between the fancier's purposeful choice and the movement of
natural selection in wild systems, he was in danger of anthro-
pomorphizing the process. Darwin's reference to selection's
"invisible hand," influenced by Adam Smith's economic language, didn't help matters, and he never quite did his part to
clarify his language well enough to prevent such misconcep-
tions.

But Darwin did, with his lovely pigeons, accomplish what
he'd hoped. He shook his naturalist colleagues out of the cozy
insularity that kept them from seeing any connection between
their own pursuits and the study of domestic animals. He
turned the tables on them by using the fancy pigeons, an entire group of beings dismissed by science, to force the possibility of diversity's springing forth from an original unity. It

was only one imaginative step, now grounded in a clear analogy, from the enclosed dovecote to the wild aviary of the
Galápagos, and of the earth itself.

There are two things (besides the resulting evolutionary
exegesis) that make Darwin's involvement with the pigeon
fancy both beautiful and singular. The first is its insistent re-
belliousness. Darwin thumbed his nose at the supposed parameters of the naturalist's ken. He traversed, with glee, a
boundary clearly marked in both the social and scientific
sands. At the same time, he brought good scientific study
right into the heart of the household, tangling his work up
with the study, the kitchen, the garden, the children. While his
barnacle studies were carried out from the home as well, they
were more of a house*bound* activity, not the sparkling, social,
feathery sort of work he grew into with his pigeons. Here is a
lively example of the ways in which a natural education can be
wrapped, as it eventually must be, about the daily life of home
and family. And while there may have been some social haughtiness involved in Darwin's time with the fancy, it clearly dove-
tailed an honest devotion, a kind of love. In a time when it
was socially acceptable for a gentleman to publicly bestow affection upon only two manners of beings — horses and
dogs — Darwin could, at the oddest times of day, be found
stroking the necks of his adored Almond Tumblers and whispering (who knows what?) into their feather-covered ears.

CHAPTER TWELVE

God and the Nightingales

Happy are those who find wisdom, and those who get understanding. . . . Her ways are ways of pleasantness, and all her paths are peace. She is a tree of life to those who lay hold of her; those who hold her fast are called happy.*

— *Book of Proverbs*

"Pure white with gray trim, like a very beautiful pigeon." This was my husband's good response to my morning query, "What does a nightingale look like?" for I realized that, lacking nightingales in North America, most of us have no idea. Was the nightingale as big as a pigeon? "A bit smaller," Tom assured me, while shaving, "but of course I don't really know. I don't know why I think that." That's another curious thing. Many people have a vague inner sense

---

*The words of this chapter of the Book of Proverbs were adapted for the anthem sung during Darwin's funeral at Westminster Abbey.

that they do know what a nightingale looks like, even though they don't. "A streaked breast, orange on its head," my sister, a science teacher, suggested. "A dark bird with a yellow throat," offered my brilliant friend Maureen. I had studied nightingales in field guides years before I saw one, so I knew, intellectually, what they were supposed to look like. Still, I was surprised to see one in the flesh. In spite of the guidebook illustrations, I had managed to secret away another nightingale image in the corners of my mind. Somehow I expected the long-tailed aqua-and-pink bird of Asian-inspired porcelain to appear right there in the gray Kentish countryside.

A nightingale is, like our common robin, a kind of thrush. Both birds have the same shape, but nightingales are somewhat smaller, more slender, and drably plumed. Their backs are pale brown, and their bellies creamy beige, with a warm wash of brown across the breast. I got my first really good look at a nightingale quite recently, wandering the Down House gardens, the estate where Darwin spent the last forty years of his life. The gardener, planting Victorian heirloom potatoes in the large plot near Darwin's greenhouses, assured me that the warm sun shining that April day was an anomaly, that exactly one week earlier it had been snowing. I'd been in London just a couple of days and took him at his word. "That's a nightingale, yes?" I inquired about the loud, bubbling, flutey voice that kept rising from some invisible singer among the bushes. He looked at me oddly for a moment. I suppose it was strange for someone with binoculars as nice as mine to be asking about such a common bird. "Yes, of course," he politely recovered. Later I would come back to learn more about nineteenth-century heirloom potatoes, but first I went in search of the bird, which, with its ventriloqual voice, was slightly more difficult to find than I expected. But there it was, soon enough, surrounded by shrubby branches and staring at me with a very large black eye, waiting for me to leave before singing again, and then again.

Late in their lives, Charles and Emma Darwin would go nightingaling in the evenings, walking the paths around their home, wandering off into the woodland-edged fields, eventually finding a lovely, grassy spot to lie back and watch the birds in flight as the voices of perching nightingales floated around in the trees. Nightingales are pragmatic flyers — competent, but not particularly elegant. They do not wheel and arc gracefully as swallows do but, like most thrushes, fly straight from the place where they are to the place they are going. Still, humans across cultures tend to mingle nightingales, and the resonant song issuing from their small, plain bodies, with poetry. It is a sensibility that predates Keats's famous ode, and it seems that the poor boy was responding to, rather than creating, a quality of the nightingale when he leaned his sickly body against the plum tree on his Hampstead grounds and put pen to paper.

The image of Charles and Emma among the nightingales is a comforting one. Even as Darwin was writing his memoir, in which he would claim to have lost all poetic sensitivity, he was spending his twilight hours wrapped in a kind of contentment, actually watching birds without any pretense to experimentation, or even collection of facts. Just watching and resting.

I've already mentioned several times that I find Darwin's autobiography to be a funny little book. I first read it about twenty years ago, when I was studying the *Origin* and hoping to round out my understanding of the theory with some insight into the man. At the time, I took the memoir at face value. It's charming in the classic Darwinian fashion, typically self-effacing, and it expresses a beautiful, innate wonder before the natural world that complements a more purely scientific investigation into the factual underpinnings of various theories. "My love of natural science," Darwin wrote, "has been steady and ardent." I still enjoy reading the autobiography now and then, preferably outdoors and out loud, either to myself or to my little daughter, who has taken to Darwin as a sort of uncle. But in the intervening years, studying Darwin's notebooks, diaries, and correspondence, I came to realize the extent to which Darwin's account is incomplete, a version of himself that, while it may be in line with the image that, late in life, he hoped to pass on, bears only occasional resemblance to the whole of the lived life.

In an odd way, this is one of the things that I love about Darwin's autobiography. The moments that have been embellished, refined, or entirely left out are just as revealing as those that are plainly told. The memoir is alternately funny, insightful, sad, and shockingly oblivious, and while it is a wonderful book, it cannot be taken on its own. Still, perhaps because it was penned in his maturity, it is often regarded as the definitive statement of Darwin's interior life. Nowhere is this more apparent, or more misleading, than in the scientific establishment's orthodox stand regarding Darwin's theological position.

In the *Autobiography*, Darwin repudiated any connection to traditional Christian faith and offered an actual refutation of Anglican creedal belief, grounded in Humean skepticism. Might the "grand conclusions" of traditional religious belief "be the result of the connection between cause and effect which strikes us as a necessary one, but probably depends merely on inherited experience?" Darwin's doubts were fitting, and even necessary, if his vision was to be consistent.

Darwin was working around the entrenched nineteenth-century worldview that portrayed humans as entirely separate from the animal world — as different from animals as animals themselves are from plants. Humans alone stood at the peak of creation. Humans alone were created in the image of God. Humans alone possessed a soul. While we humans in the early twenty-first century certainly behave as if we are the center of the universe, and while a majority, if pressed, might express belief in a kind of anthropic principle (the notion that evolutionary history has been busily striving toward a particular goal — Us), we do not typically think deeply enough about these biases to be aware of them as overt philosophical principles. But in Darwin's time there was a steadfast and comforting orthodoxy, grounding society and religion, that the human place in creation was both evident and ordained.

This sensibility was complemented by an ingrained trust in the continued intervention of a Creator God in the everyday movement of life. Though Newton's natural laws were allowed to prevail *alongside* God (who was, by virtue of astrophysics and gravity, no longer required to spin the planets and hold everything down lest it spiral off into space), supernatural intercession was still central to an understanding of earthly, biological life, where Genesis explained that the earth was created just after the sky, day, and night, and just before vegetation; where special creation governed the species makeup of the planet; where all species were always the same, and Linnaeus's codified scientific naming simply clarified Adam and Eve's without implying any biological fluidity now or ever; and where bad behavior might be punished by earthquakes, illness, or an unlucky game of cricket. These concepts — the special, separate supremacy of humankind, and the continued supernatural intervention of an anthropomorphized male Creator God — formed the very fabric of traditional belief, and they were scarcely questioned, not by working people of any class, not by the gentry or the aristocracy, and not by the majority of the academic clergy. Nearly all accepted social and religious structures twined to uphold these artifices.

Poor Darwin understood, fearfully at first, but more bravely as the decades passed, that he was required to tear all of this to shreds. If there was any kind of deity involved, then it worked through and alongside natural earthly laws, such as natural selection in the evolutionary process, not by continuous, supernatural intervention. And humans were netted to the whole of earthly life, wondrous and endowed with unique capabilities, to be sure, but not existing in privileged separation from the rest of creation, or even — Darwin was clear on this — as its pinnacle (notice his title *The Descent of Man,* not the *Ascent,* recognizing human reliance on past and existing life, rather than the reverse). In his notebook, Darwin linked this view to his characteristic humility regarding the human place in the natural world. "Man in his arrogance thinks himself a great work worthy [of] the interposition of the deity. More humble and I believe truer to consider him created from animals." Not even bothering to wipe off his hands after digging for worms, Darwin grasped the nineteenth-century social fabric firmly and ripped it straight down the middle. He was an agnostic, he concluded in his memoir. Yes, he liked the word very much, and capitalized it: an Agnostic.

It is this word, more than any other, that has been exultantly seized upon by those who would either criticize Darwin for his pure atheistic materialism or turn to him in order to justify theirs. But in either case, Darwin's intent has been distorted. The word itself is seized, rather than Darwin's own more softening context, which centered on the limits of human knowledge instead of on an outright rejection of engagement with theological questions. "I cannot pretend to throw the least light on such abstruse problems," Darwin wrote. "The mystery of the beginning of all things is insoluble by us; and I for one must be content to remain an Agnostic." The term *agnostic* was coined by Darwin's friend and defender Thomas Henry Huxley in 1869, so when Darwin applied it to himself in his memoir, the word was only a few years old. Huxley used it in a particular way, trying to mark his own path through a changed world in which he had lost his old faith without being quite sure of a new one. He was an *a-gnostic*, a non-knower, one who could not claim to know ultimate matters with rational certainty or through the empirical processes of modern science. Huxley himself could not claim such understanding, and neither, he believed, could anyone else. He disdained the arrogance of popular theologians who maintained they could explain the ways of God in the world, but he was likewise critical of atheists who scoffed at the significance of theological questioning altogether. This was, to Huxley, another kind of arrogance, an assumption that because humans cannot know something, it follows that it doesn't exist. Epistemologically, agnosticism offered an engaging, but humbler, middle way.

It wasn't until the century after Darwin that the ordinary understanding of the word *agnostic* came to carry more negative, antireligion undertones. As the prominent English philosopher of religion William E. Phipps writes in his thoroughly researched book, *Darwin's Religious Odyssey*, "In practice, the style of life of contemporary agnostics is often the same as that of an atheist, and, unlike Darwin, they often enjoy pummeling religion." Both Huxley and Darwin were careful to separate their agnosticism from atheism. Darwin stated many times over in his correspondence and notebooks that he had "never been an atheist." He also insisted that it was possible to be a theist, in fact an "ardent theist" (here he was thinking of his friend the American botanist Asa Gray), and still subscribe fully to his own scientific views. At the time, atheism was most often mentioned in the same breath as the dialectical materialism of Marx and Engels — a full reduction of all processes to the material condition — a position that Darwin found philosophically fraught. Darwin did refer to himself, half-jokingly, as a materialist in an early notebook ("Oh you materialist!" he reveled gleefully, setting himself against established thinking). That is another moment, taken out of context, that has fueled the supposed connection of Darwinism to philosophical materialism. But in another note, Darwin clarified, "By materialism, I mean, merely the intimate connection of kind of thought with form of brain." Phipps explains: "The term pertained to his reflection on how psychological traits originate and was unrelated to atheistic metaphysical materialism."

It was not theism per se, but the unthinking distortions of modern theism, that Darwin opposed. He was particularly provoked by those who, though they resided in the highest niches of academe, insisted on projecting an anthropomorphized image of their deity, as if God were "a man, somewhat cleverer than us." Naturally, this God was kind enough to have specifically arranged creation to provide for our every need.

Darwin was incredulous that the Reverend William Whewell, a philosopher of some depth, could have actually suggested that the length of night had been determined by God to coincide exactly with the human need for sleep. Or that another scientist had concluded that God created plants to control soil erosion for the benefit of human agriculture. He stood perplexed before the arguments from design that insisted on reducing the powers of the supposed Creator to the imaginative limitations of finite human minds.

In his *Natural Theology*, William Paley's flagship illustration for his argument supporting the theological necessity of design was the human eye. He carefully outlined the anatomical intricacy of the optical nerves, the delicate functioning of the iris, the sensitivity of the pupil. He compared the eye with the very best telescopes of the day, noting that the pinnacle of modern optics paled in comparison with the complexity of a single oculus. It was obvious that, just as the telescope was manufactured by a learned maker, the eye was fashioned by a Creator of far greater skill, power, and insight.

None of this made a shred of sense to Darwin, and thinking about it, which he did extensively, tied his already sick gastrointestinal system into even more knots. Why were these theists, men like Paley whose intelligence was clear and whom Darwin sincerely admired, so quick to diminish their Creator by insisting that he worked in any way analogous to human processes. "May not this inference be presumptuous? Have we any right to assume that the Creator works by intellectual powers like those of man?" If there was some kind of divine basis for life, then surely it was beyond simple human reasoning. Darwin's understanding was more in line with the biblical verses of Isaiah, in which the divine voice whispers, "Your ways are not my ways, your thoughts are not my thoughts."

As he grew older, Darwin kept pressing his life against the minutiae of earthly existence, gathering his data, reading, thinking, and seeing in his homely little facts, over and over, the movement of the whole. The more he learned in this way, the more, like Socrates, he realized he did not know. In his youth, there may have been moments when he thought science might hold the key to infinite understanding, but there were no such moments in Darwin's mature years. Rather, his life-long study of natural things allowed him to draw a slender perimeter around his knowledge, and to stand within it, quiet before the things of which he, in human finitude, must remain ignorant. He recognized that this perimeter was not imperme-able; the questions on either side often infused one another, and even Darwin's most academic treatises exposed these con-nections. Darwin was humble in the presence of mystery, yes, but he was more than humble. He was uncommonly com-fortable.

In all of the lovely meticulousness of the *Origin*, in the heaps of fact and detail, the movement of Darwin's argument, which is really quite elegant and not horribly complicated, can get a bit tangled. Basically, Darwin notes that nature's fecun-dity is everywhere observable (think of the mother quail that shows up in spring with fourteen fluffy chicks), and far more offspring are produced than any natural place can possibly support (each time the quail wanders through our yard, there are fewer and fewer chicks until, by late summer, there is only one or, if the parent birds have been both alert and very fortu-nate, two). Darwin learned from his work with animal breed-ers and taxonomists that no two individuals are alike, and so, he inferred, in the ensuing struggle for survival, individuals of a given population will differ in their probability of survival.

Darwin called this differential survival rate natural selection, and since many of the differences between organisms are heritable, natural selection will, over the course of many generations, result in evolution.

Darwin employs just one illustration in *The Origin of Species*, a branching diagram that demonstrates the diversification of species, in a section with a label that, even for Darwin, is long: *The Probable Effects of the Action of Natural Selection Through Divergence of Character and Extinction, on the Descendants of a Common Ancestor.* Darwin penned a pretty summary of the chapter:

> As buds give rise by growth to fresh buds, and these, if vigorous, branch out and overtop on all sides many a feebler branch, so by generation I believe it has been with the great Tree of Life, which fills with its dead and broken branches the crust of the earth, and covers the surface with its ever-branching and beautiful ramifications.

The Tree of Life is a positive spiritual metaphor spanning many traditions, including the Judeo-Christian, where it bookends the Bible, appearing as a connection to eternal life in both the paradisiacal garden of Genesis and as a promise to saints in Revelation. Darwin deftly connects the poetic power of the image to his radiating biological vision, rooted in the past and pressing toward the future.

Darwin's Tree is really more of a bush. It is trunkless, and implies an infinite branching. It is certainly not growing upward into the textbook pyramid structure we so often see, with humans balanced neatly, if precariously, on top. Rather, though Darwin avoids the problem of human origins directly in this book and doesn't label his branches, the thinking reader could not help grasping Darwin's purpose. Humans are thrown right into the branchy tangle, alongside apes, frogs, flounder, snails, and moss. The writing is lively in this section of the *Origin*, as Darwin clearly loves his Tree and all of its "beautiful ramifications."

Darwin's use of the word beautiful here is singular. He is critical of the common interpretation of natural beauty — that the colors and shapes of flowers, their sweet smells, the iridescent feathers of the hummingbird were all created for human enjoyment and edification. Fully aware that color and fragrance most often exist in the natural world to attract insect pollinators or mates, Darwin nevertheless reserves a place for the word beautiful, one that he does not take lightly. Darwin has lost a beloved daughter. He has witnessed the seeming ruthlessness of natural life, built as it is upon death after death. He has, as a particularly sensitive human, never come to terms with the suffering of earthly beings, and in spite of his claim to agnosticism, there is one thing he feels sure that he knows: Earthly suffering is not caused by the manipulations of a judging Creator. No, given what he has seen, he will not belittle the struggle of earthly existence in this way. Instead, he draws on an innate sense of awe in the face of natural truth. His vision encompasses all of this, and understands it, all of it, as beautiful. Beautiful in its color, in the twining of its breadth and minutiae, in its liveliness, its loveliness, its quiet, its suffering that never ends, its moments of earthen delight, which, likewise, never end. "Immense and wonderful," Darwin says of wild life. All of it is suspended in an expansive goodness, in a beauty that is, by the simple grace of its existence, redeemed within the naturalist's vision.

In Darwin's thinking, both the past and the future converge in the living beings before us, all of which have grown out of a sustaining history and will participate, all of them, by the uniqueness of their deaths and their lives, in the creation of the future. Darwin senses the simultaneous strength and frailty of such a perspective and finds within it a natural graciousness. At the end of the *Origin*, he stresses the dignity of such a vision when compared with special creation. "When I view all beings," he writes, "not as special creations, but as the lineal descendants of some few beings which lived long before the first bed of the Cambrian system was deposited, they seem to me to become ennobled."

In the next, and last, paragraph, Darwin invites the reader into his process. "It is interesting to contemplate a tangled bank," he says, drawing us into the naturalist's task, "clothed with plants of many kinds, with birds singing on the bushes, with various insects flitting about, and with worms crawling through the damp earth, and to reflect that these elaborately constructed forms, so different from each other, and dependent upon each other in so complex a manner, have all been produced by laws acting around us." To my mind, it is not the conclusion of this musing, but its unfolding, that is most significant: *It is interesting to contemplate a tangled bank.* If we are to come to understand the workings of life, then something is asked of us, and it has little to do with reading big books, even if Darwin wrote them. We are to sit still, to become quiet, to spend time in the presence of small living things, to watch, to listen, to reflect. Darwin concludes:

> There is grandeur in this view of life, with its several powers, having been originally breathed by the Creator into a few forms or into one; and that, whilst this planet has gone cycling on according to the fixed law of gravity, from so

simple a beginning endless forms most beautiful and most wonderful have been, and are being evolved.

Darwin knows what he is doing here, drawing suggestively on the biblical metaphor of an original "breath," maintaining the open-endedness of myth without lapsing into literalism. In part, he is attempting to appease his religiously minded critics. But he is doing more than this. He is revealing the breadth of his bold earthen vision, turning as it does about these creatures "most beautiful and most wonderful." In these few lines he elevates the material world to the stuff of the naturalist's faith, claiming a new kind of worth for his path of natural attunement.

The common thinking in religious circles is that theology has challenged, and continues to challenge, Darwinian thought. But a more thoughtful outlook might be one that allows Darwinism to challenge theology and keeps us, even when evolutionary truth is honored, from belittling either biology or theology by muttering superficially to our children, "Yes, honey, evolution is *how* God created the universe." The best theology will engage deeply with questions of life and science and will meet them, one day, with a conception of divine presence that is expansive enough to allow the full, chaotic play of cosmological evolution. The Georgetown theologian John Haught believes strongly that Darwin's challenge to theology "may prove to be not so much peril as gift." In his thoughtful book *God After Darwin*, Haught writes, "Evolutionary biology not only allows theology to enlarge its sense of God's creativity by extending it over measureless eons of time; it also gives comparable magnitude to our sense of the divine participation in life's long and often tormented journey."

William Phipps has suggested that the panentheistic outlook might complement a Darwinian science. The term *panentheistic*, meaning "all is *in* God," is very different from the
similar-sounding *pantheism* — "all *is* God." Panentheism refers
to a ground of existence "in whom we live and move and have
our being." It represents a middle way between pantheism,
where there is no room for relationship, as all is subsumed
into a great One, and a God that remains totally transcendent,
aloof from creation. It is possible that Darwin may have
found some resonance in such a view, and it does seem to echo
some of his comments, though it is doubtful that any theological articulation would have persuaded him to give up his
shielding agnosticism.

In his memoir, penned more than fifteen years after the
first publication of the *Origin*, Darwin "retracts" his youthful
experience in the Brazilian rain forest, where he sat over-
whelmed by the sense that "man is more than the mere breath
in his body." Now, he writes, "the grandest scenes would not
cause any such convictions and feelings to rise in my mind."
Further, he "cannot see that such inward convictions and feelings are of any weight as evidence of what really exists." He re-
nounces his early conversion in one fell swoop that I long
regarded as stingy.

But I see it differently now. Darwin was writing late in life.
He had lived in one place for forty years, going almost
nowhere. He wasn't falling into raptures, and had not for some
time, not for decades. He wasn't spilling open beneath tower-
ing rain forest palms. He was just standing in one place for
hours, watching worms spin the soil. ("If only he had something to *do*," remarked a faithful gardener, "I am sure he would
be much better.") Darwin was perfecting a quieter reverence.

Early in the *Beagle* journey he had come to understand his own smallness, the fact that his precious individuality mattered only insofar as it was wended about a wider existence. In his eyes, both this wild breadth and the unique creatures it cradled became worthy of his best attention, his "ardent" love, as he called it. In the Brazilian rain forests he experienced his own conversion, from a student to a pilgrim, opening himself in a rush of mystical connection, then working to assimilate such a change into the everyday work of a journeying naturalist. But later in life, fifty years later, these experiences seemed heady and youthful, as in fact they were. Darwin's mature reverence did not require travel, or novelty, or never-seen beings, or trees very much taller than his house. It required nothing more than himself, and his family, and the place where he lived. Reading Darwin's early notebooks, his *Ornithological Notes*, and other works, I am always intrigued with the specificity of his observations, the details he bothered to record that so often pass beyond our everyday notice. But these were nothing next to what he eventually accomplished by going nowhere. Darwin knew every inch of earth that surrounded Down House, and what might be found there. He knew every bird and its song, its comings and goings. He knew many of them not as representatives of species but as individuals — the nightingale with the slightly longer bill, the one with the missing toenail, the one with the odd warble at the end of its morning song. He knew where they were, even while he was at work in his office or lying in his bed, sick. He knew them, as it were, by heart. "There is grandeur in this view of life," he wrote. It was enough.

His body had been a constant disappointment to him for all but his early adulthood. By the time he was in his fifties, ill health was almost a permanent state for Darwin, emphasized

by his increasingly hunched
posture and the unkempt tangle of his gray beard. He was
plagued with dizziness, giddi-
ness, eczema, weakness, sleep-
lessness, exhaustion, and an
array of severe gastrointestinal
symptoms. Much of Darwin's
perceived lack of social grace
can be explained by embarrassment over his chronic flatu-
lence, which forced him to
avoid invitations and to aban-

DARWIN'S SANDWALK

don dinner parties early. Modern specialists continue to debate the cause or causes of Darwin's complaints. While some
still hold to the theory, modern archival records regarding
Darwin's health appear to refute the idea that his later troubles
stemmed from intestinal parasites or from Chagas' disease,
picked up on the *Beagle* journey. The going hypotheses involve
various combinations of nervous psychological tendencies, di-
verticulitis, gallbladder troubles, a slew of potential gastrointestinal disorders, and a measure of hypochondria.

But in spite of his various ailments, Darwin was content.
With his independent wealth he was able to pursue his love of
science single-mindedly. He was a good, affectionate father,
and he enjoyed a remarkably felicitous marriage with the indulgent Emma. The two shared a genuine fondness for each
other, a quiet love that deepened through the grief, sickness,
and joys of a long marriage. For years on end, Darwin maintained a daily routine that rarely varied. He breakfasted,
worked, strolled, lunched, read the papers on the sofa, wrote
letters in the study, listened to Emma read to him, strolled

again, worked, rested, ate, played backgammon with Emma, listened to her play the piano, and retired. The walks were essential — rain or shine, Darwin and Polly, his loyal fox terrier, would make thoughtful perambulations around the woodsy sand walk that lay on the edge of the grounds. This thinking path became, more than any other place, Darwin's spiritual residence, a place of renewal, sustenance, contemplation, and delight. As he continued to age, Darwin kept puttering around his gardens, alternately poking the soil and watching the birds, until the very last days of his life. He rambled about in his invalid's body, his energy gradually declining, until he went to bed, had a heart attack, and died. He was seventy-three — a good, ripe age for his time.

I suppose stories of Darwin's deathbed conversion were inevitable. The best known is based on the report of a dubious-sounding character named Lady Hope, who claims to have visited Darwin prior to his death and to have been told, by Darwin himself, that he had embraced Christian biblical truth and was deeply troubled that he had caused so many good people to question their own faith. He repented. One of Darwin's most prominent biographers, James Moore, researched the myth extensively and found that Lady Hope did in fact exist and that she does seem to have visited Darwin about six months before he died. Her story, he says, has the ring of truth to the extent that her details of Darwin's home, life, and even conversational style seem authentic. But there is no way to corroborate her highly unlikely tale, which grew into a genuine legend, complete with a vision of Darwin clutching a Bible at the moment of death, renouncing evolution, and begging for a clergyman. Darwin's family, though, say they heard him whisper minutes before his passing, "I am not in the least afraid to die," the words of a man who, lacking faith in traditional immortality, lay content in the knowledge of a full life lived well. I like to think, too, that although science cannot teach one to die well, as Darwin seems to have done, a life attuned to the cycles of nature can offer such lessons, and in this Darwin was schooled daily. He died embraced by his own sense of earthen goodness, tired and unafraid.

One of the seemingly odd things about Darwin's mature years, and one that appears to have caused him some distress, was his inability to enjoy the arts as he once had. He mentions the higher sentiments that music, and particularly poetry, once incited in him, dimensions of himself that he felt became entirely closed off to him in the last decades of his life. He suggests that it was his habit of mind, always whirling about scientific facts and ideas, that caused him to lose his aesthetic facility. Ease in any subject must be cultivated, he thought, or it will be lost. Darwin's health was also exceedingly poor, and higher feelings are always more difficult to access when one is nauseous or riddled with headaches. Whatever the causes, I find myself wishing that Darwin had been easier on himself, and that he could have seen his own life's work as deeply related to the matter of poetry in some way.

Darwin would not have been aware of Emily Dickinson's poetry, or even her existence, though they were contemporaries of a kind. Dickinson died just four years after Darwin, an ocean away. Only a handful of her poems were published during her lifetime, and most of these were passed on to various editors by her correspondents, not by Dickinson herself. Her distinctive punctuation and capitalizations were "corrected" for public consumption. I often imagine the two of them, Charles and Emily, in the same breath. Both lived, by choice, as recluses (being a woman in the nineteenth century, Dickinson is often characterized as a spinster, but her solitude was far stronger and more deliberate than that commonly associated with the term *spinsterhood*). Both wrote from imaginations teeming with ideas that they assumed, rightly, the public was not prepared to welcome. Both worked for decades on perfecting the expression of their thoughts. Both wrapped their writing in ribbons, closed it up in drawers, and left it there for years. Both possessed a call to life's work that revolved around the themes of life, death, time, eternity, and nature. Both could not answer to traditional religious expression, and both created their own. I don't remember how I used to read Dickinson's poem "I died for Beauty" back in high school, before I knew much about Darwin. But now I can't help reading it as a dialogue between the two of them, a kind of resolution to Darwin's poetic foundering.

> *I died for Beauty — but was scarce*
> *Adjusted in the Tomb*
> *When One who died for Truth, was lain*
> *In an adjoining Room*
>
> *He questioned softly, "Why I failed"?*
> *"For Beauty", I replied —*
> *"and I — for Truth — Themselves are One —*
> *We Bretheren are", He said —*
>
> *And so, as Kinsmen, met a Night —*
> *We talked between the Rooms —*
> *Until the Moss had reached our lips —*
> *And covered up — Our names —*

If Darwin could have looked down and seen this same blending moss on his own green feet, perhaps he would have perceived the poems he himself sprinkled in his path, the naturalist's path, where science and art, truth and beauty, nightingales and the divine wend about one another, talking into the night.

CHAPTER THIRTEEN

The Naturalist's Faith

On the favorable side of the balance, I think that I am
superior to the common run of men in noticing
things which easily escape attention, and in observing
them carefully.

—*Autobiography*

My favorite room at Down House is Darwin's study,
arranged precisely by historians after a period
photograph, with a great deal of the original furniture, books, tools, and notes. In the diary Darwin kept on the
Beagle, he had referred happily to his little cabin, with all its
bottles, books, papers, and leaf pressings stored so tightly and
neatly, as his "intricate corner." It is a happy, self-satisfied reference, and I thought of it when I saw Darwin's room. He
maintained some of his seafaring tendencies throughout his
life, particularly in the way he stowed his research — bits of
paper, feathers, rocks, and barnacles were tucked into small

wooden drawers and files that he had made to line the walls of his office.

DARWIN'S STUDY

The central table is now arranged very much as it must have been on one of Darwin's typical workdays. There are several stacks of books, including a couple of large scientific volumes torn straight down the spine, as was Darwin's habit when he felt a book was troublesomely large. There are stacks of notes in his curling, increasingly unintelligible script. There are scattered bones, shells, feathers, skulls, and bird skins. And there are several of Darwin's tools — worn brass tweezers, dissecting implements, magnifiers, microscopes that he had remade to suit his purposes. Even in Darwin's day none of it was new. None of it was state of the art. None of it resembled in any way the scientific labs in the city, where his colleagues daily labored. His ruler was a ratty piece of cardboard with the increments marked by hand in pen — inaccurately, it turned out. For a scientific thinker who had achieved such stature, Darwin's tools, and the greenhouse where he studied his orchids and carnivorous plants, seem almost inappropriate — simple, amateurish, and odd. Where he had once fancied the very best of scientific toys, Darwin matured into a naturalist who preferred, in fact prided himself upon, making do with what he had. The result was a winsome, makeshift tangle that, I cannot but think, represented Darwin's own interior mind rather well.

For all our attempts to unravel Darwin's life and mind, the development of his theories, and the chronology of his thought, the most stunning aspect of Darwin's work as a naturalist is its simplicity. William Beebe calls him the most "natural of naturalists," referring to his ability to derive his famous science entirely from observation and meditation upon the lives, habits, and homes of living organisms. Darwin never lost his watcher's faith that the smallest happenings might be of import. Much that he found noteworthy would not be considered "statistically significant" today, would not find entry into the mathematical models that characterize modern ecological science, and would certainly not meet the criteria of controlled replicability required by reputable biological studies. Much of Darwin's most important work would be confined to the scientific rubbish heap labeled "anecdote." Yet it is from his string of stories, attended over a lifetime, that Darwin's elegant theory of natural selection, and all that it implies, was gathered.

In a recent issue of *Natural History,* an article by T. V. Rajan appeared, titled "Would Darwin Get a Grant Today?" Rajan laments that the study of "whole organisms" in natural settings is becoming antiquated as enthusiasm for molecular biology gains power. The naturalist's endeavor, he writes, "has increasingly been perceived as a poor cousin to molecular biology and has even, in some scientific circles, been contemptuously dismissed as stamp collecting." While field biology is still an active field, it no longer claims the fundamental place in ecological study. Darwin himself would have been thrilled by molecular studies, by the myriad ways they both justify and improve his own work. But he would never have misplaced the organism itself in examining either its ecosystem or its DNA.

Increasingly, it seems science is allowed to encroach upon the most basic human decisions — what to eat, what to buy, what to teach our children, what to abhor, what (if anything) to protect, what (if anything) to protest. And at the same time, those educated in the biological sciences are separated more and more from their subject. Today, a student of ecology can earn her Ph.D. without ever leaving the university campus, working entirely with statistics and mathematical models on her desktop computer. At the same time, the language through which we come to understand the natural world

DARWIN'S GREENHOUSE

has been steadily migrating out of the hands of the general public into the realm of pure academia, where journal articles, even those about subjects that might be readily understood by a knowledgeable amateur, are encoded in the secret language of the academic scientist. This is not a terrible thing in and of itself. The harm comes in allowing the complete handing over of biological understanding to the realm of science alone.

Scientific language is necessarily lexiconic, seeking a pure clarity. At its best, science writing attains a kind of elegance. But the language of academic science must, in order to succeed, banish the unspoken, eliminating the hidden spaces between words. Yet it is here, in the spaces between what can be seen and what can be spoken, that the naturalist's faith so often lies. This is why it may be called faith. Intimacy, residence, patience, a sense of dwelling alongside wild nature, earthen insight, gratitude, affection, kindness, a kind of grace, a kind of joy — all of these unutterable things find a place in the naturalist's task.

The well-known Cornell biologist Thomas Eisner was

responding to the academic marginalization of natural history, and the felt need to reclaim a proper role for the naturalist's understanding, when he created a new class called "The Naturalist's Way." The class is lecture-based, featuring some of the nation's very best thinkers on the whole range of the subject, placing artists, nature writers, and poets alongside ornithologists, atmospheric scientists, entomologists, and, yes, even molecular biologists ("visionary" ones, Eisner says; we could call them "molecular naturalists"). Eisner's hope is not to be dismissive in any way of molecular science, which, he writes, "adds shape, form, and depth to the inquiries that, so far, have been driven by the natural historian." Rather, the goal is to "restore the glamour of classical natural history, let students know that it is still very much alive, and provide reassurance that the questions natural history asks are as vital as ever." The class has been enthusiastically received by a population of young people who, in addition to being educated as scientists, hope to turn to nature study for a sense of stability, a way of seeing, a way of living.

It is not my purpose here to propose a new model for science or scientific education (though I am glad there are thinkers engaged in such questions). Instead, I want to suggest that if we desire accurate knowledge about the creation of which we are a part, we should not leave the cultivation of such knowledge in the hands of science alone. We need to participate, at a most basic level, in our own biological education. We do not need to become scientists in order to do it, but we do need to become attentive and involved. There are hundreds of ways, in the ordinary dailiness of human life, to enter the naturalist's faith.

On our kitchen table lies a large notebook with a hard cover, a spiral binding, and white pages. It is always open.

Anyone may write it in anytime, but we usually begin dinner, just before we join hands to give thanks for our meal, with a nod in its direction. Does anyone have any phenological observations to record? Phenology, the study of cyclical phenomena in the natural world, has become a family practice. We are attempting to pay careful attention to the migrations of birds, the unfolding of leaves, the blooming of flowers, the arrivals of butterflies, the outbreaks of tent caterpillars, the singing of warblers, the nesting of crows, the cycles of weather, the night sky, the interplay of all these things about our household as well as in the forested parks and on the Puget Sound beaches that we frequent around our West Seattle home.

The benefits of such a simple practice are palpable. Much more than the simple naming of organisms, or even a study of their habits, this phenological awareness immerses us in the rhythmic dimensions of the natural world. It is enhanced by a study of ecology and ethology but is more personal somehow, even visceral. While watching for moments that we might bring to one another's attention at our little evening ritual, we become engaged, watchful, and increasingly resident. Certainly we do not confuse our small urban yard and its four chickens with the wild, but in our observations we do become aware of the presence of wildness, even here, and it increases our outward sense of connectedness to the true wild places, within our Cascadian bioregion and beyond. And when we do come to such places, we are softened and attuned, we seem to slough off our urban layers more readily and willingly.

A phenological notebook has little meaning if it is one year old, and even at three years old, like ours, it should still be considered new. Its treasures spiral and layer with the years, and with the comparisons from one year to another — the first Wilson's Warbler song, the day the crows began to hatch, the one and only Band-tailed Pigeon we ever saw in our yard, the rare winter night we could see Saturn's rings with a spotting scope. I have learned that there is phenological software available, suitable for the backyard naturalist, that will collate all of this information. If you type in the date of the first warbler's arrival, say, the computer program will automatically place this information next to the warbler arrival dates you typed in for past years. I realize that some people are tickled by such technological neatness, and if it brings them joy, then so be it. But both the Luddite and the poet in me wants to hang garlic around all of our doorways to ward such things off. I believe that the visceral act of recording our observations is one more dimension of immersing ourselves in the cycles we attend, and in flipping through the pages, seeking for ourselves the date of last year's first robin, and the years' before that, we find not only the robin but the movement of time, the movement of life, and the events that surrounded that particular event, that clustered around it like a constellation.

More than this, our notes are accompanied by essential trimmings that no software could possibly accommodate — streaks of mud, coffee stains, newspaper clippings, taped-in feathers, bits of eggshell, the perfect wing of a mourning cloak butterfly, and myriad little sketches. My own drawings are unimpressive (though most of them are better than Darwin's), but my daughter's are terrific. Yesterday I was greeted by a giant and extremely happy butterfly — its entire body, one big circle, was a happy face. It had purple antennas and four good wings, each striped with black and yellow. "It's a Tiger Swallowtail," Claire informed me, trying to be cool but clearly pleased with her work. "I saw three of them today."

At the same time that such observations heighten our sense of relatedness, they also, perhaps ironically, heighten our sense of superfluity. The rhythms of nature simultaneously sustain, contain, transcend, and ignore us. The more-than-human nature of the world can be a broadside. Watching the liveliness of a crow in windy flight, we sometimes find it difficult to remember, even as we are uplifted, that the crow's exuberance exists entirely without reference to human recognition.

I spoke recently with the fabulous naturalist Dennis Paulson, director of the University of Puget Sound's natural history museum, and asked him what a person hoping to grow as a naturalist ought to do. "Go outside and study nature," he told me enthusiastically. "Read. And travel — travel is so important." Surely travel intensifies and broadens our understanding of what is possible in wild nature, but I think its primary value for the naturalist may lie in the way it enhances our understanding of our home places, making our knowledge more specific, and more honest. I believe the deep understanding of a particular place, the place where we live, is essential. Just as the study of a single species of bird might lead to an understanding of an entire forest, so the deep commitment to learning the wild, natural cycles of a particular place can lift us into a broader natural insight of our home ecosystem, our bioregion, our earth. In his introduction to the poet Holderlein's *Hymns and Fragments*, Richard Sieburth wrote about the Greek notion of *omphalos*, the idea that "all the holy places of the earth come together in a single place." Sieburth writes, "To be truly here is to be everywhere — any locus is potentially an *omphalos*." It's always a bit startling to remember that while Darwin gathered much of his initial inspiration in the tropics, for the last forty years of his life he didn't go anywhere. He walked his sand walk, circled about on the strange wheeled chair in his mind-office, and grew so close to his home landscape that he very nearly sank into the earth itself.

Darwin's combination of biological attunement and obsessive record keeping made him a natural phenologist. But even Darwin could not have kept up in this regard with his American contemporary, Henry David Thoreau (who had read Darwin's *Origin of Species* gratefully, as a kind of revelation). In his last years, Thoreau became a maniacal phenologist, recording particularly the cycles of tree growth in the Maine woods, where he lived. His records lie close to his radical philosophy, stressing immediate involvement with wildness. On the inside cover of our phenology notebook, I have inscribed his words as an homage and a reminder: "Talk of mysteries! — Think of our life in nature, — daily to be shown matter, to come in contact with it, — rocks, trees, wind on our cheeks! the *solid* earth! the *actual* world! the *common sense! Contact! Contact!*"

Darwin called himself a "philosophical naturalist," both looking and thinking deeply. But here in the early twenty-first century, when any pimple-faced day camp baby-sitter calls himself a "naturalist," it is a word worth reclaiming. In the nineteenth century, one who studied earthly life with reasonable depth and intelligence could claim naturalist status. But Darwin and his colleagues did not know the ecological degradation that we face today, though the seeds were surely being sown in their time.

When Darwin was in the Galápagos, he was assured by the governor that enormous tortoises, ones that six men could scarcely lift, were once common. But already in the 1830s such animals were rare, and the tortoise population overall was no-

ticeably dwindling. "Mr Lawson thinks there is yet left sufficient for 20 years," Darwin wrote, in an early recognition of
the human capacity to decimate an entire population of nonhuman beings. Even with this insight, Darwin would be stag-
gered by the human-wrought changes in the modern animal
landscape.

The delicate tracings of the fossil record teach that over
the course of a typical millennium we may expect about one
avian species to go extinct. Wild nature in the evolutionary
context is changing, unfolding. This is as it should be. But in
recent millennia this earthen pace has been interrupted. One
hundred twenty-eight bird species have gone extinct in the last
five hundred years, 103 of these since 1800 — more than fifty
times the expected "background rate," as it is called. The organization Bird Conservation International currently predicts
that by 2100, 460 more species of birds are highly likely to
have become extinct. The statistics for other classes of animals
are similar, and because the evolutionary stories of species are
so intricately connected, many biologists worry that the current extinction crisis will have ramifications far beyond the
lineages of particular lost species. The unfolding of the evolutionary process itself is increasingly at risk.

Globally, there are 1,186 threatened birds, and 99 percent
of these are in jeopardy because of human activity. Habitat
degradation in all its guises — development, agricultural conversion, cattle grazing, logging, mining, draining of wetlands,
pollution — is the primary cause. Other risks to birds are ex-
acerbated by the loss of habitat, including hunting, persecution of raptors, impacts of invasive species, capture for the pet
bird trade, and even slingshotting by children. The extinction
crisis facing birds and all other forms of wildlife is a global one,
but vulnerable species do cluster in particular places. More than

90 percent of the extinctions since 1800 have occurred on islands, where geographic ranges are limited and endemism runs
high. The new-world tropics, with their stunning degree of
avian diversity within rapidly disappearing moist forests, hold
high concentrations of threatened species. Many of the birds
Darwin studied in South America are in precipitous decline,
including a small tinamou called the Lesser Northura, the Yellow Cardinal, the Saffron-cowled Blackbird, various spinetails,
the Eskimo Curlew, the Greater and Lesser Rheas, the Galápagos Hawk, and the flightless Galápagos Cormorant, to call
just a very few by name.

It is difficult to know what to do with such information,
such a list. In 1954, Rachel Carson confided to a roomful of
women journalists her belief that "the more clearly we can
focus our attention on the wonders and realities of the universe about us, the less taste we shall have for destruction."
This was long before it was fashionable to say such things, and
I think of Carson often as I attempt to bring the role of the
naturalist into proper focus. It is in light of such words that
my modern sense of a naturalist, while restoring some of the
depth that the term carried in Darwin's day, must necessarily
cut further. I believe that the naturalist's practice today most
involve both an attentive study of the biological life unique to
our geographical place and an attempt to bring our own lives
into increasingly authentic relationship with that life. It must
involve knowing our home place deeply and well enough to
live elegantly within its bounds and to speak strongly for its
needs.

The means by which we come to this knowledge need not
be fancy. It entails the simple, daily, practical work of treating
animals, trees, insects, and plants — as well as the myriad
foodstuffs, homes, and tools created from them — with care,

respect, and as gracious a measure of knowledge as we can muster over time. I know two or three living humans whom I think of as true naturalists. I consider myself a stumbling naturalist-in-progress, with flagrant shortcomings. Becoming a naturalist is a lifelong process, involving constant forgiveness, good humor, radical joy, and no sensible timeline. It means standing, as Emily Dickinson wrote, with our "souls ajar," knowing that patience and faith have as much as science to do with this insight we are tending.

In cultivating such faith, I will turn to Darwin's good, plain, eccentric, sincere, struggling, brilliant, and humble writings again and again in my life. He reminds us, as he painstakingly learned himself, that we, too, are animals, connected to life, past and present. That we are earthly residents, with the innate capacity for attentive, authentic relationships within the sum of life as we live, work, and play at the borders of nature, science, and culture. That we become alive and embodied in our attention to life's detail. That nothing in the natural world is beneath our notice.

ACKNOWLEDGMENTS

These pages have grown out of the insight, wisdom, and encouragement of many wonderful people. Foremost among these are my literary agent, Elizabeth Wales, editor, Terry Adams, and editorial assistant, Sarah Brennan.

I am grateful to the librarians at the Seattle Public Libraries and at the University of Washington's Allen and Suzallo libraries for their constant calm assistance in matters both commonplace and arcane.

Several thinkers in the fields of ornithology and the biological sciences took time to discuss their work and views. I would particularly like to thank Dr. Dennis Paulson at the University of Puget Sound, Dr. Sievert Rohwer at the University of Washington, Dr. Tim Eisner at Cornell University, and Dr. Frank Steinheimer at the British Museum of Natural History.

For their loving, shining presence, I am ever thankful to my family — parents Jerry and Irene Haupt, parents-in-law Al and Ginny Furtwangler, and especially my husband and daughter, Tom and Claire Furtwangler. I offer a special song of praise to my sister, Kelly Haupt, and all public school science teachers like her, who face the current unnatural, pseudoreligious "creationist" wind and teach evolution as the expansive, beautiful, scientific truth that it is.

SUGGESTIONS FOR FURTHER READING

This list is eclectic, and far from complete, but it offers a few of my favorite books that relate to the many dimensions of Darwin's natural history work, and to the life of a naturalist.

BOOKS BY AND ABOUT DARWIN

Charles Darwin: Voyaging, Princeton University Press, 1996, and *Charles Darwin: The Power of Place,* Knopf, 2002, both by Janet Browne.

In my opinion, Janet Browne's two-part, twelve-hundred-page biography is the very best. Along with rich detail regarding all aspects of Darwin's life and science, there is in these books a rare, uncompromising feel for Darwin's foibles and humanity. The first volume contains the most thorough treatment of the *Beagle* years available within a broader biographical treatment.

Charles Darwin's Beagle *Diary,* edited by Richard Keynes, Cambridge University Press, 2001.

The day-to-day musings of the young Darwin throughout the *Beagle* journey, as he wished to present them to family, friends, and his future self, are wonderful reading, better than any secondhand account. Richard Keynes's good notes add to the pleasure.

The Correspondence of Charles Darwin, edited by Frederick Burkhardt and Sydney Smith, Cambridge University Press, 1985–.

Darwin was an enthusiastic correspondent, penning more than fourteen thousand letters. His extant correspondence is collected in an enormous, relatively recent ten-volume set, and what a delight it is to spend a week or more doing nothing but dipping into these books and being transported in mind and time. Cambridge also put out a very short single volume, *Charles Darwin's Letters: A Selection, 1825–1859,* also edited by Burkhardt, and with a lively forward by Stephen Jay Gould. It can hardly substitute for the whole, but it is good for the home library.

Charles Darwin's *Ornithological Notes,* edited by Emma Nora Barlow, *Bulletin of the British Museum (Natural History) Historical Series,* London, 1963.

Darwin's early development as a naturalist unfolds within these avian observations. Nora Barlow's notes are sometimes controversial but highly readable.

Darwin's Religious Odyssey, by William E. Phipps, Trinity Press, 2002.

An account of Darwin's religious and spiritual life by a respected British professor of religion and philosophy, this book explores the complex intersection of science and faith.

RELATED READING

God After Darwin: A Theology of Evolution, by John Haught, Westview Press, 1999.

Haught, a Georgetown theologian, trenchantly ponders the ways that both neo-Darwinian science and intelligent theology might challenge and deepen one another.

Owls and Other Fantasies: Poems and Essays, by Mary Oliver, Beacon Press, 2003.

Oliver is a modern poet-guide to the natural world; her poems are a blessing for the naturalist, and for all who seek a renewal of inspiration in the wild world. *Owls and Other Fantasies* is particularly birdy, but Oliver's other books are lovely, too.

Always, Rachel: The Letters of Rachel Carson and Dorothy Freeman, 1952–1964, edited by Martha Freeman, Beacon Press, 1995.

A long collection of letters between Rachel Carson and her dear friend Dorothy Freeman, whose affection for each other grew over their mutual love of nature, which they shared on beaches and woodland trails, over books, and beneath hair dryers, and which spanned years and miles. These letters are reminders of the many ways that attention to the processes of nature can inform and sustain almost every aspect of daily life.

A Feeling for the Organism: The Life and Work of Barbara McClintock, by Evelyn Fox Keller, W. H. Freeman and Company, 1983.

A gorgeous life of McClintock, who transcended the rules of both science and gender in the mid-twentieth century, bringing the naturalist's way of seeing into her studies of genetic transposition. McClintock literally watched her corn plants grow, treating each one as a separate, particular being as she teased advanced scientific insights from their colored kernels.

BIBLIOGRAPHY

Adams, Samuel. "Notes on the Rhea or South American Ostrich." *Condor* 10 (1908): 69–71.

Avian Brain Nomenclature Consortium. "Avian Brains and a New Understanding of Vertebrate Brain Evolution." *Nature Reviews: Neuroscience* 6 (2005) 151–59.

Barlow, Emma Nora. *Charles Darwin and the Voyage of the Beagle: Unpublished Letters and Notebooks.* New York: Philosophical Library, 1946.

Berry, Wendell. *Life Is a Miracle: An Essay Against Modern Superstition.* Washington, D.C.: Counterpoint, 2000.

———. *The Art of the Commonplace: The Agrarian Essays of Wendell Berry.* Edited by Norman Wirzba. Washington, D.C.: Counterpoint, 2002.

BirdLife International. *Threatened Birds of the World.* Cambridge: Lynx Editions and BirdLife International, 2000.

Browne, Janet. *Charles Darwin: Voyaging.* Princeton: Princeton University Press, 1995.

———. *Charles Darwin: The Power of Place.* New York: Alfred A. Knopf, 2002.

Codenotti, Thais L., and Fernando Alvarez. "Mating Behavior of the Male Greater Rhea." *Wilson Bulletin* (March 2001): 85–89.

Darwin, Charles. "Notes Upon the Rhea." *Proceedings of the Zoological Society of London* 5 (1837): 35–6.

———. 1839. *Journal of Researches into the Natural History and Geology of the Countries Visited During the Voyage of* H.M.S. Beagle *Around the World.*

Reprinted as *Voyage of the Beagle*. Introduced by Steve Jones. New York: Random House, 2001.

———. 1859. *The Origin of Species by Means of Natural Selection*. Reprint edition. New York: Random House, 1993.

———. *Charles Darwin's* Beagle *Diary*. Edited by Richard Darwin Keynes. Cambridge: Cambridge University Press, 1988.

———. "Darwin's Ornithological Notes." Transcribed and edited by Emma Nora Barlow. *Bulletin of the British Museum (Natural History) Historical Series* 2 (1963): 201–78.

———. *Charles Darwin's Zoology Notes and Specimen Lists from H.M.S. Beagle*. Edited by Richard Darwin Keynes. Cambridge: Cambridge University Press, 2000.

———. 1872. *The Expression of the Emotions in Man and Animals*. Reprint edition, edited by Paul Ekman. London: HarperCollins, 1998.

———. *The Autobiography of Charles Darwin, 1809–1882, With Original Omissions Restored*. Edited by Nora Barlow. London: Collins, 1958.

———. 1881. *The Formation of Vegetable Mould Through the Action of Worms, with Observations on Their Habits*. New York: New York University Press, 1990.

———. *The Correspondence of Charles Darwin*, edited by Frederick Burkhardt and Sydney Smith, Cambridge: Cambridge University Press, 1985–.

de la Peña, Martin R., and Maurice Rumboll. *Birds of Southern South America and Antarctica*. London: HarperCollins, 1998.

Dennett, Daniel C. *Darwin's Dangerous Idea: Evolution and the Meanings of Life*. New York: Touchstone, 1996.

Desmond, Adrian, and James Moore. *Darwin: The Life of a Tormented Evolutionist*. New York: W. W. Norton & Company, 1994.

Dickinson, Emily. *The Poems of Emily Dickinson*. Edited by R. W. Franklin. Cambridge, Mass.: The Belknap Press, 1998.

Farber, Paul Lawrence. "The Development of Taxidermy and the History of Ornithology." *Isis* 68 (1977): 550–66.

———. *Discovering Birds: The Emergence of Ornithology as a Scientific Discipline, 1760–1850*. Baltimore: Johns Hopkins University Press, 1997.

———. *Finding Order in Nature: The Naturalist Tradition from Linnaeus to E. O. Wilson*. Baltimore: Johns Hopkins University Press, 2000.

Feduccia, Alan. *The Origin and Evolution of Birds.* New Haven: Yale University Press, 1996.

Gould, Stephen Jay. *The Structure of Evolutionary Theory.* Cambridge, Mass.: The Belknap Press, 2002.

Grant, Peter R. *Ecology and Evolution of Darwin's Finches.* Princeton: Princeton University Press, 1986.

Griffin, Donald R. *Animal Minds: Beyond Cognition to Consciousness.* Chicago: University of Chicago Press, 2001.

Gruber, Howard E., and Valmai Gruber. "The Eye of Reason: Darwin's Development During the *Beagle* Voyage." *Isis* 53 (1962): 186–99.

Habegger, Alfred. *My Wars Are Laid Away in Books: The Life of Emily Dickinson.* New York: Random House, 2001.

Haught, John F. *God After Darwin: A Theology of Evolution.* Boulder, Colo.: Westview Press, 2000.

Hauser, Marc. *Wild Minds: What Animals Really Think.* New York: Owl Books, 2001.

Heinzel, Hermann, and Barnaby Hall. *Galápagos Diary: A Complete Guide to the Archipelago's Birdlife.* Berkeley: University of California Press, 2000.

Hölderlin, Friedrich. *Hymns and Fragments.* Translated by Richard Sieburth. Princeton: Princeton University Press, 1984.

James, William. *The Varieties of Religious Experience: A Study in Human Nature, Centenary Edition.* New York: Routledge (2002).

Katagiri, Dainin. *Returning to Silence: Zen Practice in Daily Life.* Boston: Shambhala Publications, 1988.

Keynes, Randal. *Darwin, His Daughter, and Human Evolution.* New York: Riverhead Books, 2001.

Keynes, Richard Darwin. *Fossils, Finches, and Fuegians.* Oxford: Oxford University Press, 2003.

Kricher, John. *A Neotropical Companion: An Introduction to the Animals, Plants, and Ecosystems of the New World Tropics.* Princeton: Princeton University Press, 1997.

Larson, Edward J. *Evolution's Workshop: God and Science on the Galápagos Islands.* New York: Basic Books, 2001.

Maddocks, Fiona. *Hildegard of Bingen: The Woman of Her Age.* New York: Doubleday, 2001.

Mayr, Ernst. "Darwin and Natural Selection: How Darwin May Have

Discovered his Highly Unconventional Theory." *American Scientist* 65 (1977): 321–27.

Paulson, Dennis. "The Need for Continued Scientific Collecting." 2003. Unpublished.

Phipps, William E. *Darwin's Religious Odyssey.* Harrisburg: Trinity Press, 2002.

Remsen, J. V. "The Importance of Continued Collecting of Bird Specimens to Ornithology and Bird Conservation." *Bird Conservation International* 5 (1995): 145–80.

Ridgely, Robert S., and Guy Tudor. *The Birds of South America. Volume I: The Oscine Passerines.* Austin: University of Texas Press, 1989.

————. *The Birds of South America. Volume 2: The Suboscine Passerines.* Austin: University of Texas Press, 1994.

Schultz, Stewart T. *The Northwest Coast: A Natural History.* Portland, Oreg.: Timber Press, 1990.

Secord, James A. "Nature's Fancy: Charles Darwin and the Breeding of Pigeons." *Isis* 72 (1981): 163–86.

Snyder, Gary. *Earth House Hold.* New York: New Directions Publishing, 1969.

Steinheimer, Frank D. "Charles Darwin's Bird Collection and Ornithological Knowledge During the Voyage of *H.M.S. Beagle,* 1831–1836." *Journal of Ornithology* 145 (2004): 300–320.

Stepniewski, Andrew. *The Birds of Yakima County, Washington.* Distributed by the Yakima Valley Audubon Society, Yakima, Washington, 1999.

Sulloway, Frank J. "Darwin and His Finches: The Evolution of a Legend." *Journal of the History of Biology* 15 (spring 1982): 1–53.

————. "Darwin's Conversion: The *Beagle* Voyage and Its Aftermath." *Journal of the History of Biology* 15 (fall 1982): 325–96.

————. "Further Remarks on Darwin's Spelling Habits and the Dating of *Beagle* Voyage Manuscripts." *Journal of the History of Biology* 6 (fall 1983): 361–90.

Turner, Jack. *The Abstract Wild.* Tucson: University of Arizona Press, 1996.

Weiner, Jonathan. *The Beak of the Finch: A Story of Evolution in Our Time.* New York: Alfred A. Knopf, 1994.

White, Michael, and John Gribbin. *Darwin: A Life in Science.* New York: Plume, 1997.

Lyanda Lynn Haupt holds a master's degree in environmental ethics and philosophy, with a personal emphasis in ornithology, and the emerging science of conservation biology. She is the author of *Rare Encounters with Ordinary Birds*. Her articles and reviews have appeared in *Image, Open Spaces, Wild Earth Journal, Birdwatcher's Digest,* and *The Prairie Naturalist*. She lives in Seattle.